Introduction

I had two grandmothers and three grandfathers.   I know almost nothing about them.   I am sure that they had many interesting experiences, and did creative things that are lost to posterity.

I am writing this for my Grandchildren, and their Children, and their Children's Children. I have had many interesting things happen to me, and I have been blessed with a bit of creativity.

> Come with me to the Vertical Assembly Building at NASA, Cape Canaveral, ride to the top in an elevator immediately adjacent to a space rocket, and peer down into the astronaut's capsule.

> Experience the survival from a 220 Volt shock while standing in water.

> Ride in the cockpit with me and look over the stern of our sailboat to watch the approach of sixteen foot waves.

> Share with me the design road for a device that continues to be the standard of an industry after fifty years.

> Examine three U S patents.

> Come ride with me to see the Big Sky of Wyoming in the Tetons from the back of a horse.

I use this book to do a bit of teaching. Read on and learn a bit of the history of the Mississippi river. Learn about the rotation of the earth, and how it interacts with the sun. Read on, and learn about the mysteries of the ocean depths, and how to explore them using SCUBA gear safely. Maybe you can learn a bit about Heat Transfer, and Mechanics of Materials.

Read on and learn of love for family.

Mechanical Memoirs
And related stories

This is December 2011, and I am 75 years old. I have just received a pacemaker, and I am expecting some very good years ahead. Grandson Sam recently asked me about my experience with patents. I relayed to him some of the things that I had done, but it was difficult to be thorough on the telephone. So, I am going to write about some of the noteworthy mechanical things that I have been involved with over the years, and some of the stories that are memorable to me.

The earliest thing that I can remember was World War 11.

When I was yet a tot, the war was going on. I didn't know why there was a war. I remember that there was rationing of many things that were said to be important to the war effort; Nylon, Tires, Gasoline and a lot more.

My dad was not drafted into the army because he was a father, and had skills as a Mechanical Engineer. My Dad's part was to be a 'black out' Warden. At that time the Government was worried about an attack from Japan or Germany by airplanes on a bombing mission. The idea was that if major metropolitan areas were to put out all of their lights, that the enemy would not be able to distinguish us as a city. These days, with GPS, there would be no doubt. Well, my dad, Carl, had to police an area of several blocks to see that all of the lights were out during the designated time for a black out. We lived at 3504 45th Avenue South in Minneapolis MN.

We listened to the 'war news' daily on the radio. TV had not yet been invented. I was four or five at the time, and had little comprehension of the matter. My Brother George and I were busy digging, playing and eating.

I remember that we dug a 'fox hole' in the boulevard between the sidewalk and the street. We had seen pictures of men at battle in foxholes shooting and defending themselves. Our foxhole was about two feet wide, and two feet deep. We used wooden guns to defend it.

Tom's   fox hole   circa 1941

I remember misbehaving. I was rarely spanked, but one day, for some inexplicable reason, I decided to play 'trucks' in the middle of the intersection. I didn't realize that the cars had to swerve around me. My dad arrived home to see me at play that day. Well, he grabbed me by the arm and hauled me to our yard. He gave me a very hard spanking on the rear end. I remembered that spanking for a long time, and did not play in the street again. Was that a lesson, or was that 'child abuse'?

Pa

Parents are often good mentors, and can show us through example how to excel in some specialty. That can be the Arts, Medicine, Sports, The Humanities, and etc   My Dad liked Mechanical stuff, and I watched him do amazing things with his hands.

In 1946 after the big War, Pa was working for the Standard Oil Company. He had a '46 Ford Club Coupe that was his company car, and used it to call on his customers. He wanted a family car. He was able to get a 'deal' on a 1939 La Salle. The La Salle was made by General Motors, and was the little brother to the Cadillac. The La Salle was big and black and had an eight-cylinder in-line engine. Well, this car had a bad engine because it was not maintained very well. The owner probably skipped a lot of oil changes. So, Pa took it upon himself to overhaul the engine in our little garage at 2711 Garfield St N.E. Minneapolis MN.

He took the engine all apart, and laid the parts out in an organized fashion so that he could reassemble it correctly at a later date. At 10 years old, I served as a very minor helper. We washed all of the greasy parts in Naphtha. He then examined the parts to know which needed replacement or fine-tuning. There were pistons, compression rings at the top of each piston, and oil rings below the compression rings. The pistons had wrist pins to connect the piston to the con rods that led to the crankshaft where journal bearings cushioned and lubricated the con rod/crankshaft connection. There were large bearings and seals at each end of the crankshaft to imbed it into the engine block. Above the engine block was a head that held the valves that were used to admit fuel/air mixture, or to exhaust the products of combustion. The head also had a spark plug for each cylinder. The crankshaft was connected outside of the engine to the transmission where the gears changed the RPM of the shaft that connected the whole array to the rear end. The rear end transferred the torque from the drive shaft to the axles, and thence to the wheels. There also was an array of auxiliary devices the enable the engine to properly do its job of converting gasoline to rotational power in the tires. These devices included the Carburetor, Fuel Tank, Fuel Pump Distributor, Generator, Six-volt starter, Water Pump, Radiator and Exhaust system. In those days there was no such device as Power Steering, Power Brakes, Air Conditioning, Electric Windows, or CD player.

Well Pa discovered that the cylinders were scored, and had to be bored out. That meant that he had to buy over size pistons and new rings. All of the bearings and seals were replaced. The head was ground to be once again flat, and the valves were ground and re seated. There might have been other things that he did, but the most amazing thing was that he was able to put the whole thing back together. The engine worked as good as new when he was done. It did not burn oil, and had plenty of power. Pa gave the car a very nice wash job, and coat of paste wax. He then took the whole family out for a spin in our nearly new car. Wow.

Pa taught me another secret. Changing the brake pads on a shoe and drum brake assembly can be difficult because of the many small parts, springs, and etc. Taking it apart, and buying new pads is the simple part. Putting it back together was not so simple. Pa's solution was to take a Polaroid picture of the assembly before disassembly. He used that photo in reassembly.

It was that experience, and watching Pa tackle any thing in the house that needed fixing from Radiators to Furnaces to woodwork that sowed those first seeds in me to be inquisitive, observant, and constantly looking for ways to make things better. Pa even converted a front porch to be a part of the living room by tearing out the front divider between the living room and the porch and then insulating, painting, and carpeting the new bigger living room. He installed a 'Picture Window' which over looked the front lawn with its two 30 ft spruce trees at each side of the lot.

2711 Garfield St NE Minneapolis MN   circa 2012

More about Pa

At about the age of seven or eight, I lived in the above house, and knew the neighborhood very well. I went to St Charles Borromeo Catholic grade school, and was honored to be a "Police Boy". In that assignment, I was to direct traffic near the school so that the pupils could cross the street safely. I also was in the Boy Scouts, played basketball, and served as an altar boy.

One day, when I was in the drug store on Johnson Street, I spied a dime that had been left on the counter. No one was around, and no one seemed to own the dime, so I picked it up and put it into my pocket. Somehow, when I got home, or perhaps later that day, Pa saw the dime, and asked me where I got it. I told him about the drugstore counter. He explained to me that someone had probably left the dime after he or she took a newspaper or a candy bar, and that that dime was the property of the drug store. Uh Oh.

Pa took me by the hand, and we walked the three blocks to the drug store. I wondered why we didn't take the car, but now in retrospect, Pa wanted a little time to lapse so he could talk to me about honesty.

When we got to the drug store, Pa had me go in and ask for the Pharmacist. I gave that dime back to the Pharmacist, explaining that I had made a mistake.

Now, that lesson has lasted me a whole lifetime. Every time I've had an opportunity to take something that wasn't mine I remembered the walk to the drug store. I have always felt very good for doing the right thing. Thank you Pa.

So, thanks to Pa and Ma for helping to shape my young life. I had an 'I can do anything' attitude.

Pa

Boat Number One

Our family took vacations at Minnesota lakes. Boating was always a big deal. My grandparents had a cottage on Deer Lake in Wisconsin near Taylors Falls. Here is a pic of me at age 4 holding on to that great big Neptune.

Somehow, at age 13 or 14, I got the very strong desire to build a boat. To finance this project, Georgie, my brother, and I started delivering newspapers a couple of years previously. Now, let me tell you, delivering papers in Minnesota during the winter is not for wimps. I remember Sunday mornings when we left home at 5 AM in the dark and pulled a toboggan loaded with 120 thick Sunday newspapers through the snow in temperatures that were well below Zero. We were cold. The paper route gave each of a little financial independence. My savings account might have been a couple of hundred dollars when the boat building bug bit.

I did some research through popular science magazines, and found a small, fast, simple and inexpensive boat plan. It was an 8 ft. racing hydroplane. It would be the right size to assemble in our little garage. I bought the plans to build it. This little boat was built from one sheet of 4 x 8 x ¼"marine plywood, some ¾" Sitka spruce for the frames and stringers, a ¾" Oak transom, 1 ½" Oak motor board, and some canvas to cover the deck.

The shape of the bow was pram like and about 3 ft across. The bottom was about 30" wide, and had chines that abutted the bottom, and angled upward at about 45 degrees to meet the deck structure. The keel was 1' x 1 ½" Oak. To get some shape to the bottom, a very narrow vee was cut from the front of the bottom plywood and toward the aft about 2 ft. The front of the vee was about 3" wide. There were about 6 frames of ¾ x 3 spruce and screwed together with slotted head brass screws and glue made of 'weldwood powder'. I used a brace and bit to drive those screws home. I learned that if the screws had a dab of glue on the tip, they would go in a little easier.

Now, each of these frames was made about 6 " too high so that when they were fastened to the keel and turned upside down, they sat on the garage floor and held the upside down frame in a position where it could be easily worked on. Stringers of $\frac{3}{4}$ spruce were screwed to each frame at the top and bottom of each chine, and intermediate on the bottom. These stringers were the bedding for the relatively light $\frac{1}{4}$" bottom. With a solid framework setting on the garage floor, I cut the vee in the front of the bottom, and began fastening it to the framework using brass screws and glue. As I worked forward, the $\frac{1}{4}$" plywood resisted bending, so I soaked the plywood for a few days using regularly dampened towels over the area that was to be nicely formed into a curve. Then, with the plywood still wet, I screwed it to the framework working toward the front. As I approached the front, the 'vee' that was cut began to close up to compensate for the curvature. The chines, which were about 10" wide, but curved to fit up with the bottom, and the deck line, were installed. Now the boat was turned over, and the over length frames were cut off to make way for the decking. Deck stringers were installed on either side of the cockpit to allow for 6" side decks to be placed on both sides of the cockpit. On the bow, more stringers were added to support the canvas decking. The canvas decking was fastened to the stringers using copper flat head short nails.

The little boat was finished off by adding the motor board to the transom, a steering wheel, and a fin on the bottom for stability. With Pa's help, I purchased an eight and one half horsepower Champion Motor. I installed a home made steering bar of aluminum to the motor so that the steering cables could attach to it. The steering cables passed through pulleys to a drum behind the dashboard. The drum was on the same shaft as the steering wheel.

Now the fun began. We took the boat to Lake Minnetonka atop Pa's car and on a carry rack. I installed the motor and gas tank. I put on a life jacket, and started her up. Wow, that first trip across the water was very exhilarating. The bow of the boat started to flop up and down. The problem was that the angle of the motor was not set correctly, and as I accelerated, the motor pushed the stern of the boat down and hence the bow up. Well, no big deal. Pa and I adjusted the motor angle a couple of times until it was just right. Now my little boat flew across the water on an even keel, and at a speed of about 30 mph. This was much fun, and I had a tremendous feeling of accomplishment knowing I had built it myself.

We later found that I could take one passenger along if he or she sat carefully on the front decking.

This boat wasn't really a racer. It was fast enough for me, but at that time class 'A' racing hydroplanes used souped up motors, and traveled at speeds of about 60 mph.

Boat # 1

Boats #2 and #3

Well, of course, that first boat wasn't big enough. That is the way boating goes. We always want something bigger or faster. Lucky for me, we moved to a very nice place on West Arm of Lake Minnetonka. It was a 2-acre lot with a lot of frontage on the lake, and one side was bounded by a canal going into the next bay. Our folks bought it after a fire destroyed the previous house. As I recall they only paid about $5,000 for the lot. Pa designed the house. It was a good home for many years. They couldn't afford to build a garage at first. I remember that a few years later, Pa gave up cigars to finance a garage. It seems that he smoked about 5 Dutch Masters per day, and that cost about $35 per month. The loan on the new garage was $35 per month. Pa figured that the cigar company paid for his garage.

Boat #2 was built using the same construction techniques as #1, except that this time I fibreglassed the underwater seams. #2 was 12 ft long, and equipped with an old 16 hp Johnson "Green Head" motor. It was a big improvement over #1 and could carry one or two passengers, and the best of all; it could PULL A GUY ON WATER SKIS. It seemed like we were always fiddling with that old motor. The carburetor, spark, or something else was always wrong. Pa was right there to show the way for fixing.

Boat #3 was built because #2 wasn't fast enough or big enough. I was about 17 at the time, and the boat was a 16 footer, with a wide transom. It was a really good ski boat for kids. It had a mid deck with a steering wheel, could carry four or five people, and best of all. water skiing was no problem. It was a bit slow with more than one passenger or one skier. The old 16 hp motor was a problem.

The time was 1957, and Johnson motors came out with the 35 hp motor. It was the biggest available at the time (now they go up to about 225 hp). We eyed those motors with envy, and finally Pa, George and I all pitched in to buy one for about $400. Oh happy day. That motor was ideally suited to boat #3. The boat could now carry six passengers, and best of all, it could pull 3 water skiers (we all weighed about 150 lbs. at the time) The best speeds were when only one person was on board. However the rule in those days was that if you pulled skier, you had to have a second person aboard as an observer. Later that was changed when good rear view mirrors were available. Also, the rules in those days did not require skiers to wear life jackets.

Water Skis

We built many different water skis. It was a lot cheaper to build than to buy them, and we were always on a limited budget. George was as good a ski carpenter as I was. The first skis were 8 inches wide and about 5 ft long. Those big skis were really good for getting out of the water quickly, and they had a minimal drag on the boat. Next was slalom ski, that single ski that was spiffy because of the big rooster tail you could dig out with it. The rear of the ski was tapered down to about 3 inches wide, and fitted with a small fin on the bottom. Oh, were we hot stuff now!

Next came Turn-a-Bout Skis that were about 3 ft long, and had no fins. They could go forwards, backward, and sideways, and we could do 180, and 360 degree spins with them. One more step up the show off ladder.

After that was Shoe Skis. They were about 18 inches long, and had no fins. They required a 'starter ski to get out of the water. They were fun, but required a lot of concentration to operate. On my 70th birthday we had a big deck party here in San Diego with about 70 people. We had a Mariachi band, and very good food and drinks. As the party was winding down, I noticed something floating in the pool. Alas, it was those old shoe skis. Niece Kelly had dug them out of the boathouse on Hand Lake. It was the first time I had seen those skis in 35 years!

About Ma

Of course we built a ski jump to be anchored out in front of our house on West Arm Lake Minnetonka. That ski jump was about 4 ft wide and 4 ft high. Boy, were we hot stuff going over that jump at the amazing speed of 25 mph. Ma was beside herself with worry about someone getting hurt. She was pretty good to go along with most of the hair-brained stuff we came up with.

Ma was always supportive. She wasn't inventive or mechanical, but she took good care of 'her boys', and even tamed Pa a bit. I remember Ma taking care of Mike when he was small and had contracted Scarlet Fever. The doctors had cautioned her about the damage Scarlet Fever could do to Mike's heart. Mike wasn't to walk or play like kids normally do. It was something about the exercise raising the heart rate. Ma carried Mike from his bed to the table, and then to the couch, and back and forth to the bathroom. She did this for about three months. It must have worked, because Mike turned out to be very healthy, and he even was a star basketball player.

The best part of Ma was the care and tenderness that she showed to all of us. Who can forget the many hours that she spent at our bedside when we were sick? Who can forget the batch of chocolate chip cookies that came out of the oven with an eager young mouth waiting to gobble up a few of the hot ones? Who can forget the times that she made pies (especially Rhubarb) and had some left over piecrust that she baked on a cookie sheet. She dusted those morsels with sugar and cinnamon. They were hot, crispy, and very tasty. Who can forget the big tomato garden that she nurtured and watered daily that yielded those tomato steaks that we put between slices of bread, and slobbered with mayonnaise?

Ma had many friends, and liked to play bridge with them. After Pa died, she moved to an apartment/assisted care building in Spring Park. Life was good for her there for a few years. She spiraled down, and finally died there with Mike holding her hand, at the end. Thank you, Mike.

I won't ever forget walking beside Ma's casket at the funeral service at Our Lady of The Lake Church in Mound MN. The organist played 'How Great Thou Art' which was Ma's favorite religious song. The church was full of Ma's family and many friends. I had a very heavy heart, and tears running down my cheeks. I felt that I had lost one of the finest Mothers that any person ever had.

More on Skis

The best ski device of all was not even a ski. It was a Saucer. It was 3 ft in diameter, flat, $\frac{1}{2}$ inch thick, and slightly sanded around the edges. It was so simple that it didn't even require paint. The Saucer was pulled behind the boat at pretty slow speeds of about 15 mph. If you went too fast the Saucer would flatten out on the surface, and dig in if it hit a small wave. Slow boats could pull it. We would put our feet near the center of the Saucer, and yell, "hit it" to the boat driver. We would pop out of the water in just a few feet, and then motion the driver to slow down so the darn thing was somewhat controllable. We were able to do 180, and 360 degree spins, go backwards, and even could have two kids on it if they were very careful. The Saucer could somewhat be controlled by weighting the left side to turn left, and the right side for a right turn.

 Then the unimaginable tricks started. We would take a pass at the dock, and pick up a chair. Then sit on the chair, and do spins. If that was too dull, we stood on the chair, and did spins. Then of course, you could sit on the chair, and hold the tow-rope in your feet.
That would free up your hands for other stuff. Why not take a pass at the dock, and pick up a newspaper. Now we passed the other houses on the lake with our nautical reading room. Oh what fun. Next was a small stepladder. It just fit on the Saucer, and if one was very careful, he could crawl to the top of it, and stand on the top rung. Spins were a bit dicey from that perch, but we did a few.

Slalom ski

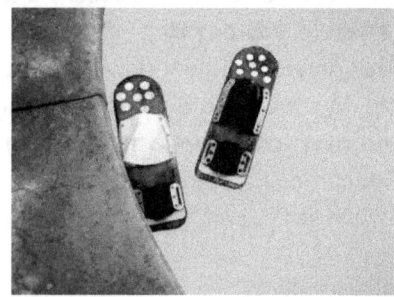

Shoe skis

House at 3143 Lakeshore Blvd, Mound MN 1960

College

I was not the best of students in High School, so when it came time to enter Engineering School, I was told that I had to go to a minor college for a year, and get at least a "B" average. OK, so I went to the College of St. Thomas in St Paul. It was a good experience, and many of the credits were transferable to the U of M.

I entered the Mechanical Engineering Department in the Institute of Technology at the University of Minnesota in the fall of 1955. This is what Pa had done a number of years earlier, and brothers George, Mike, and Jim were to do a few years later. It seems funny now that we all went the same way. I don't remember our folks ever pushing us toward any technical career. It just seemed natural for me. Engineering was then a five-year course, with 250 credit hours to be taken quarterly at the rate of about 17 per quarter.

In those days there were no computers, and calculators were just being developed. Engineering students used 'Slide Rules', a device for doing multiplication, trigonometry, and logarithms. We all carried Slide Rules, kind of like today's teens carry I phones. I still have my old Pickett Slide Rule that is now nearly 60 years old.

There were no Transistors then. Radios and some control devices used 'Vacuum Tubes'. Vacuum tubes used 'Thermionic Emission'. A grid in a vacuum was heated red hot by electricity. It then gave off electrons that were aimed at a positive grid. On the way to the grid those electrons passed through a magnetic field that slightly changed their path. The Magnetic field was driven by an exterior device. The grid collected data from the electrons, and transferred the results to circuit boards that controlled Radios, and such.

I didn't much understand electronics, and did better with Heat Transfer, Physics, and Machine design.

In my second year at the U of M, George joined me as an incoming freshman. We rented a room in an old house on Washington Avenue right across from the 'Big Ten Bar and Grill'. We had many of our friends from Mound MN with us there. The house had about ten sleeping rooms. We had a big common bathroom, but no kitchen facilities. George and I were skinny tall kids. (about 6' 4", and 150 lbs.) We kept a big jar of 'Cheese Wiz' in our room to spread on crackers when we were hungry.

We entered many of the Intramural Sports tournaments with the guys from that rooming house, and a few others. We called our teams "The Washington Avenue Bum's" We played Basketball, Football, Volleyball, and even toyed with boxing.

I remember some of the games the bunch of us pulled on one another.

We would find someone sleeping, and put a dollop of shaving cream on his hand. We then tickled his nose, and watched as the hand came up to scratch the nose. The shaving cream was then deposited all over the face. Those of us that had carried out the trick stood in the hallway, and had a good loud laugh. The guy that woke up to find the mess didn't think it was so funny. He would then plan to get a revenge trick on the perpetrators.

My favorite trick was a get even thing on Pat Guy. He was a great buddy, and always full of fun. I remembered the many tricks he pulled on me, and now was my time. I found a two-foot long live garter snake at home, and planned my revenge. I was careful not to tell anyone about it before, or after the case for fear of escalating the revenge thing. The snake was a harmless thing that fed mostly on bugs. When every one was out of the building and at class, I brought the snake to Pat's room. I pulled back the covers, and used a safety pin to attach the critter to the sheets. It didn't really hurt the snake, but it insured that the snake would stay between the sheets. When bedtime came, I made it my business to stay somewhat close to Pat's room. Well, Pat crawled into that bed, and was met by a wriggling snake. He let out a big, big whoop, and flew out of the bed. He immediately went searching for the perpetrator. We all responded to the Whoops, and had a great laugh on the guy who pulled off most of the tricks. No one knew who did it!

Engineering school was tough. It required a lot of class time, and a lot of study. George and I also worked to support ourselves. We had jobs cleaning cheese and butter machinery at Land O Lakes Creameries. We worked about 20 hours a week, and were well paid for that era. The jobs came from a contact that Pa had with the Goblirsh's next-door neighbor, who was the Manager of the Butter and Cheese packaging rooms. We kept those jobs for about 4 years.

I remember a nasty accident that happened. Quarter pound butter sticks were packaged in a 'print' machine that took 60 lb.

butter tubs, and forced the butter into a vertically flying shoe. The shoe had a slot in it the same size as the desired stick. It oscillated up and down to make the stick. At the bottom of the stroke, butter was forced into the mold, and the shoe moved upward. At the top of the stroke, another part of the machine pushed the stick out, and into the wrapper section where the butter was covered with a waxy paper. The wrapped stick then was combined with three other sticks, and put into a carton. The cartons flowed down the assembly line, and into the caser, where twenty four cartons were put into a cardboard case, and then delivered to the cooler for storage. The line was pretty well automated, but each machine had an employee watching to catch problems, and to load new supplies of wrapping materials. One day the lady operator of a print machine had stopped it to clear out a paper jam. She would jog the machine slowly to expose the problem. Well, she somehow got her finger in the slot in the flying shoe, and jogged the machine with her other hand. Very sadly, the finger was severed, and put along with butter into the wrapper. She ran quickly to the manager's office, and pandemonium broke loose. She was given a small tourniquet, and the Ambulance was called.

We stopped all of the machines in the room, and began a search for the finger. The butter stick with the finger in it had traveled down the assembly line, and neared the caser. All of us in the room opened package after package of butter until the finger was found. To this day, I can still feel the pain of that day.

Some years later the government founded OSHA, the watchdog for safety in the work place. Had the machine been properly designed, it would have been equipped with shields, and devices that would make it impossible to put ones hand in the danger area.

I had another 'memorable' event with a machine while attending the University. My machine was a 1953 Plymouth that I had bought from the Standard Oil Company's (Pa's Company) fleet when it got too old to meet their standards. I remember paying $450 for that car, and had a payment schedule with the bank that lasted two years.

Shortly after making my last payment, a buddy of mine and three or four other kids decided to go to a dance at Waconia. He had a Convertible that had a rather low center of gravity. I followed him around the curvy roads near Lake Minnetonka with two passengers. I remember the car skidding a little, and the next thing I remembered was seeing the ceiling in a hospital room. Well, fortunately none of us was badly hurt. The car however was a total loss after rolling over.

I went without a car for two years until I graduated. Georgie, my brother, was kind enough to provide transportation for me.

The Vietnam War was going on in my University years. I was a straight kid with Republican leanings, and did spend 2 years in the Air Force ROTC. I had dreams of becoming a pilot. We did a lot of marching, and I was on the rifle team. I had a right proper uniform and kept it very clean. We took training flights out of state to get us used to flying. The ROTC thing ended when we were measured for flight duty. In those days anyone over 6 ft 4 inches was not considered for fighter planes because of cockpit size limitations. I was over the limit.

Those were the last months of the War. The people of the USA were pretty fed up with all of the death and expense, and could no longer tolerate the War. The government had told us that we were in it to stop the advance of the Communists. On campus, the students were very much against the war, and many of them demonstrated against it daily around campus, and especially in front of the ROTC building. Those protests and many like them across the country were instrumental in our elected officials deciding that enough was enough in Vietnam. They brought the war to a close. Looking back, I began to despise the government for having brought us into what turned out to be a meaningless struggle that cost many lives, and many dollars. One would think that government learned a lesson, but no, the same process was to repeat itself over and over again.

Boat Number 4

While living at Lake Minnetonka, we had lots of fun. Pa had buried a 50 gallon tank near the old garage, and we had it filled with gasoline for boat use regularly. We often would spend 3 or 4 hours on a Saturday water skiing, mostly behind the 16 ft boat #3.

Then, as it will ever be, we wanted a bigger faster boat. Well, we found a 15 year old 17 foot Correct Craft inboard. It was in bad repair, so we got it at an affordable price. It had a four cylinder gasoline engine, and was fully planked in Mahogany.

I remember my buddies from grade school, high school and the University helping to put that boat back into usable condition. Bob Schwaab was working as a mechanic at the time, and he and Pa worked on the engine. They did the points and plugs thing, and I don't know what else. John Seemann, Brother George and I stripped off all of the old varnish and bottom paint. We sanded, and sanded, and sanded, and stained the mahogany back to original color. We put on many coats of varnish and bottom paint.

#4 was launched, and had that throaty exhaust heard only from a water cooled marine engine. It was a pretty thing, having been all spruced up with paint and new upholstery. It made a big wake, but wasn't too fast. Its top speed was about 25 MPH. It worked just fine as a ski boat and we used it to tour the lake.

I remember trying to 'barefoot ski' behind it. To do that I started our on regular skis, and then dropped one off, planted the one bare foot in the water, and then dropped the other ski planting both feet directly in the water. It worked OK at first. However the boat wasn't fast enough to allow my body to be held up on such a small surface area contacting the water. Shortly, I sunk a bit into the water, and soon thereafter my feet were going backward, and I landed on my chest. My feet and legs then tried to go over my head, and my back was painfully wrenched. Well, I tried that a few more times always with the same result. Alas, I was not to be a barefoot water skier.

In those days, we did a lot of dating. #4 was a good enticement for the young ladies. Now here is a story on me.

Seemann, Schwaab and I arranged a boat 'date' with four young ladies on a warm summers evening. We started out with our dates from 'Casualaire' (the name for the West arm property) and crossed over West Arm to the Lakeview. The Lakeview had a big dock, and had a dance floor with a live band, and food and drink. After a lot of dancing, we decided to go to Mack's Pizza on Smiths Bay. It was a warm moonlit night, and the trip to Mack's took only $\frac{1}{2}$ hour. The pizza was tremendous as always. We then headed back to Casualaire, through Layfayette, Crystal, and Maxwell bays. Along the way, I thought it would be good sport to fake the engine quitting. The boat stopped, but floated nicely on the moonlit water. We did a bit of Kissy facing (if you know what that is). Now, I tried to start the engine and proceed home.

Alas, the engine would not start. I coaxed and coaxed it to no avail. All we had to propel the boat was a canoe paddle. We paddled awkwardly to the nearest shore. We (the guys) stripped down to underwear and got into the shallow water. We walked that dead boat some two or three blocks to Ralph's Pure Oil station. Ralph's had a dock that I had used before. We secured the boat, and proceeded to walk about a mile back to Casualaire. All turned out okay, except that we got the young ladies home a bit late. I felt sheepish for having pulled the dumb stunt.

One evening we took the boat across West Arm (Casualaire's Bay) to a party sponsored by a friend, Jimmy Sherber who was renting there. There was a bunch of people there, and I took some of them out for a boat ride. One of those people was Cathleen Casey. The romance was started that night.

First Job

I remember My college pals and I talking about what we were going to do when we got those 'big' paychecks that Engineers get. We vowed to get a haircut every two weeks. Those days, kids with long hair were considered slops. I interviewed with a few companies. Ford Motor Company flew me to Dearborn MI to see their operation, and interview me. I remember the test facility and track there. They ran cars to death on a big oval track. They

26

did this to their own cars, and to all of the competitor's cars. After the testing, they took the cars totally apart, and spread the parts over the floor for analysis.

After a couple more interviews, I settled on a job at Land O Lakes Creameries. I was hired into the engineering section of the Dry Milk Production Department. The pay was $500 per month. We had about ten drying plants around the Midwest, each attached to a creamery. I worked under a very bright man, Reg Meade. He was tasked with making the milk drying process more efficient, and coming up with new products.

I used my heat transfer training to design tubular heat exchangers. They were chambers of cylindrical shape that had 'tube sheets' welded into each end. The tube sheet had many holes bored to receive tubes that ran from one end of the exchanger to the other. There was a space between the tube sheet end closing doors. Those doors had inlet and outlet piping all made of stainless steel. One could design the heat exchanger with varying number of tubes, and baffle them so as to get a single or multipass flow of the material to be heated (or cooled). A heating medium such as steam was introduced into the shell, but outside the tubes. The heated medium circulated through the tubes. It was possible to control the temperature of the heating medium, and to get the desired temperature increase in the heated medium by adjusting design parameters. Once a design was ready (I made the drawings), we put the design out for bid with contractors who would make the heat exchanger for us. I interacted with the contractors to see that the work was done properly.

We were also working on a method of spray drying, or fluid drying to produce a powdered whole milk that was good tasting, and easily soluable in water. One such set of experiments required a special formulization of concentrated milk to be sprayed into a Rogers Dryer. That dryer was a room about 7 foot high, 15 foot wide and 20 feet long. The spray nozzles were mounted high on one end, and the room was heated with a large volume of hot air continuously turned over. As the concentrated milk was sprayed into the room, the hot air finished the process, and the milk was powder when it hit the floor. The floor was equipped with a full width cleaner. The cleaner was many pieces of stainless steel angle iron separated by about two feet, and fastened to a chain on either end near the sidewalls. Thus, the angle irons were pulled along the floor collecting dry powdered milk. At the end of the room was a trough with a screw conveyor to remove the product to other storage. My job was to assist Reg in whatever formulation he wanted to try. Some times the drying did not take all of the water out of the particles. The damp particle then hit the floor, and clumped together with others. That made a big mess. I went into that hot room (205 degrees) many times with a big shovel to correct the mess. I had on a breathing mask, and could only stay in the room for about five minutes at a time. Aah the value of a college education!

Because of an alcohol problem, Reg got fired. He was a brilliant engineer, but had some inner need to get high on alcohol. After Reg was fired, the department was closed down.

Hyser Electric & Thiele Engineering

The time was 1962, and the government had military service a requirement for most able bodied young men. There were some exceptions: If you were going to school you could be exempt, if you were married with children, you were exempt, and if you worked in a 'Critical Skills' profession you were exempt. Well, when I was at Land O Lakes doing engineering work, I was considered to have a Critical Skill. The Country needed to have people who could contribute to the food industry. Also, men who worked in industries that provided important goods for the national effort were exempt as having Critical Skill. The government reasoned that it was better to keep the movers and shakers of industry at home. Those men with little or no skill would be drafted into the military service. Years later, that system was abandoned as being discriminatory. Still later, the draft system was abandoned in favor of voluntary service.

So now here I was without a job. I received notice that I was now classified at 1 – A. That meant that I was up for the next round of draft into the Military.

I used the Minnesota State Employment Service to look for a job. I was referred to Hyser Electric, and got hired. Hyser made electrical power distribution apparatus and packaging machinery. It was considered a 'critical industry'. I got the job just in time to avoid the military. My boss at Hyser had to fill out paper work to satisfy the Draft Board.

While working at Hyser I was married. It was a very big moment in my life. All of our friends and relatives attended a ceremony at

St Laurence Catholic Church. A reception in a nice restaurant followed, and then more partying at Casualaire. I was very happy with my new bride, my new family, and a glorious wedding day.

We moved into a small second story apartment in south Minneapolis and started a family.

I remember very well the birth of our first baby boy. He was born prematurely. I wasn't allowed to be in the birthing room, or to hold the little guy. I could only look at him in an incubator from the hallway through a plate glass window. I had a very new realization of the meaning of life. It was overwhelming, and seemed to be a big swelling in my chest. I was very happy.

But alas, Baby Boy Olson only lived a day or two. I went from being on top of the world to having the weight of the world crush me. My poor wife, Cathleen, suffered along with me.

I made arrangements with a mortuary to bury the little guy. The mortuary man came with a little white box. We took the little coffin to the cemetery. Tears flowed from my eyes often and heavily.

Hyser was a small company with only 7 or 8 employees. My job was to do anything required to get orders for our shop, and to guide the production of those products. Most of my work was clerical, and included answering the phone, and yes, even sweeping the floor.

Sometimes we would get work that required me to design machines. Whoopee.

Now let's get the credits straight. Many times, I was the main guy in the design and development process. However every time, there were others that contributed ideas, and hard work. Someone had to identify the need for something new, and it was not always me. Someone supervised my work, and let me have as much freedom as I required, but also contributed ideas. Many colleagues, and shop people contributed ideas and hard work. All of these people are necessary in the design process. I hereby acknowledge all of them.

Hyser Electric Stage Lift

Theatrical and other stages often require large things such as backdrops, curtains, and scenery to be raised off of the stage at the appropriate time, and to be lowered back at the appropriate time.

We designed an electrically powered five-chambered spool that would haul in some 1/8 inch cable which was u-bolted to the top of the curtain. When a remote button in the wings was pushed, the spool would begin to turn, and the five cables would hoist the curtain upward. At a predetermined spot, the electric would stop and the spool would come to rest. The top, and bottom locations, was set by a rotary switch that cut or started electricity to the motor through a starter. The rotary switches could be adjusted to achieve vertical positioning. The motor, rotary switches and spool were mounted in an aluminum casting, and very compact. The five chambered spool was also an aluminum casting. Oh what fun to watch it when it first did its job!

A Mechanical Computer

In the year 1962, the computer was just being developed, but was not available to us. Thiele Engineering (a sister Company to Hyser) had developed a series of 'Coupon Placers. These machines would take a coupon, small packet, or a cardboard tab, remove it from the storage bin and place it into any container that the customer wanted. Typically coupons were placed in boxes that passed in front of it on a production line.

Pillsbury Mills contracted with us to insert Burpee Flower Seed packets into cake flour boxes on the filling line. It was a springtime promotion for Pillsbury to get households to buy the cake mix in order to get the free seeds. To further complicate things, Pillsbury had six different varieties of Burpee seeds. The idea was to get each home to go back to the store for another package of cake mix, and then get a new variety of flower. Daisies, Petunias, Marigolds and etc.

The filling machine that we were to drop the flower seeds into was located in a mill in Springfield Illinois. The filling machine was called a 'Never Stop' machine. That filling machine was about 40 feet long and 10 feet wide. Flour was dropped in one end of it at the filler head from a floor above. The machine opened flat carton material, sealed the bottom, and inserted a sanitary liner. The assembled empty box then traveled to the filling head. As the box traveled around the head end it was filled by rotary filler that turned at the same speed as the box on the rotary table. The height of the flour material in the box was measured by a photo cell. If the box wasn't filled properly, an ejection solenoid kicked it off of the line and onto the floor.

This was happening at high speed. The boxes traveled down the line at about 10 miles per hour. Our task was to place a packet of flower seeds in each box, and insure that each box had a packet. Thiele designed a five headed coupon placer that dropped packets into the empty boxes before they got to the filler head. The machine would drop, pick a new batch of five packets, and drop them into the very next five boxes going down the line.

My job was to design a system that would remember if the seed drop had occurred and if it didn't, to send a signal to the ejection station requiring the ejection of the faulty box.

No small task, right.

My boss, Leavitt Anderson, contributed most of the electrical circuit design.

We positioned a photo switch directly above each of five stations that the coupon placer was to service. The coupon placer would take a packet from the storage bin using a vacuum cup. The vacuum was cut at exactly the time that the target box was at the station. The packet would drop through the photo switch field (five at a time to the correct station), and into the flour box. If the drop was successful, no signal was sent to the relay that operated the solenoid that was to trip a dog on the Memory wheel as it went by. If there was no successful drop, the photo switch would pass the information to the relay which in turn energized the solenoid corresponding to that station on the memory wheel that was exactly 20 stations from the ejection station. The memory wheel and coupon placer were synced to the Never Stop machine by a chain drive from the main power shaft of the Never Stop machine.

The memory wheel turned, and brought the tripped dog to the read out position. The read out micro switch sensed that the dog was tripped, and passed an electrical signal to the ejection device on the Never Stop machine. The time on the memory wheel between the dog being tripped, and the micro switch reading the bad news, was exactly the same as the time it took the box to travel from the offending coupon placer head and the ejection station.

The design and building of the memory device took place in Minneapolis at Hyser Electric. We mated the memory device to the Thiele five-headed coupon placer.

In designing packaging machinery, there are three different designs phases to get to the final product. The first is the design that comes off of the drawing board. The machine is then made in the shop, and low and behold it is not quite right. Adjustments are made with the help of the designers, and the shop people. The machine is tested in the shop and modified until it appears to be right. The third phase of design happens when the machine is installed at the customer's site. Adjustments are again made to adapt the design to real life on some assembly line.

I was the technical guy who accompanied the machine to the customer's site in Springfield Illinois.

Let me tell you that the first time we turned everything on after carefully mating our machine to the Never Stop filling machine. All hell broke loose. As it happened the Pillsbury plant Manager was there when we gave it the first go. The Never Stop machine, the coupon placer, and our memory wheel all started simultaneously.

As might be expected all was not in perfect synchronization, and every single box of flour that approached the ejection station was ejected on the floor. It took on a minute or so to get a pile of ejected boxes about five feet high. The accompanying cloud of flour dust was something to behold.

I was scared, and wondered if we would ever be able to pull off this deal. The Plant Manager walked away undoubtedly with some very unpleasant thoughts in mind.

It took about two weeks to get all of the bugs worked out, and that completes the third phase of packaging machine design.

Here are some crude sketches of what was going on:

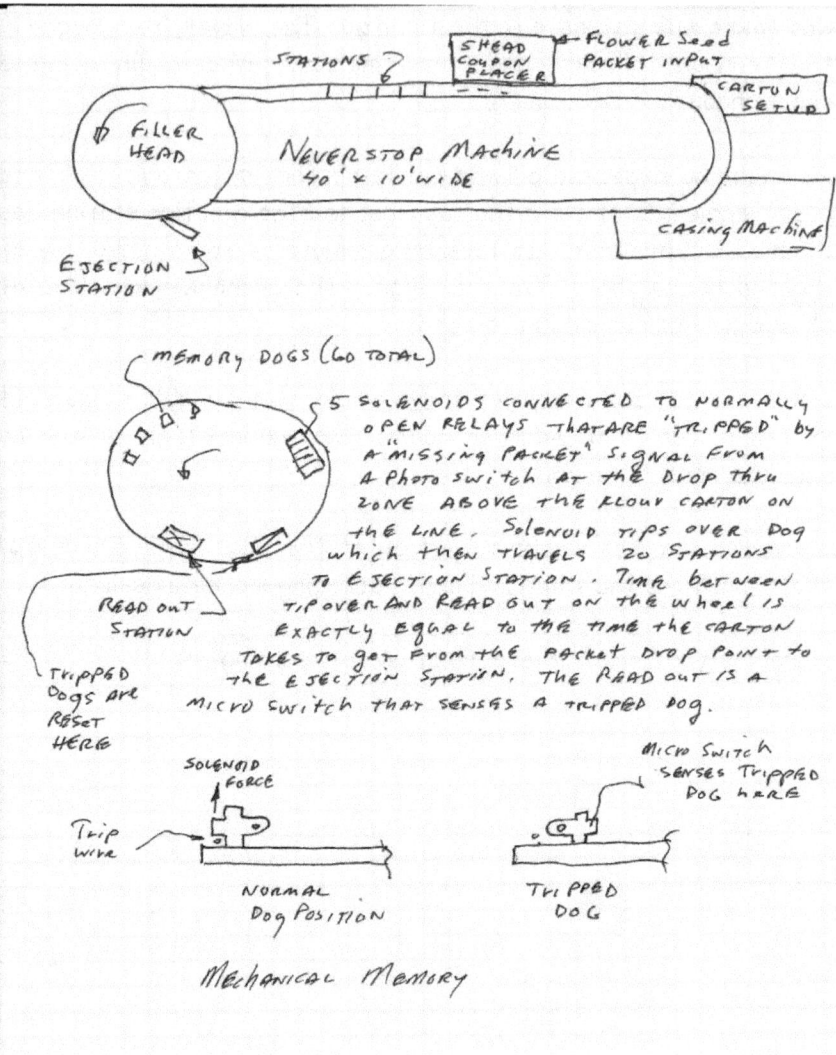

STATIONS

SHEAD COUPON PLACER

FLOWER Seed PACKET INPUT

CARTON SETUP

FILLER HEAD

NEVER STOP MACHINE
40' x 10' WIDE

CASING MACHINE

EJECTION STATION

MEMORY DOGS (60 TOTAL)

5 SOLENOIDS CONNECTED TO NORMALLY OPEN RELAYS THAT ARE "TRIPPED" by A "MISSING PACKET" SIGNAL FROM A PHOTO SWITCH AT THE DROP THRU ZONE ABOVE THE FLOUR CARTON ON THE LINE. SOLENOID TIPS OVER DOG WHICH THEN TRAVELS 20 STATIONS TO EJECTION STATION. TIME BETWEEN TIP OVER AND READ OUT ON THE WHEEL IS EXACTLY EQUAL TO THE TIME THE CARTON TAKES TO GET FROM THE PACKET DROP POINT TO THE EJECTION STATION. THE READ OUT IS A MICRO SWITCH THAT SENSES A TRIPPED DOG.

READ OUT STATION

TRIPPED DOGS ARE RESET HERE

SOLENOID FORCE

MICRO SWITCH SENSES TRIPPED DOG HERE

Trip WIRE

NORMAL DOG POSITION

TRIPPED DOG

MECHANICAL MEMORY

36

This Pillsbury plant was one of the largest powdered food product plants in the United States. It was built in 1929. When I was there in 1962, it operated with 'Man Lifts' to move people between floors. The building was eight stories high and had about 6 Man Lifts located at strategic points. Each Man Lift was eight stories high. They consisted of an eight story high conveyor belt about 16 inches wide and $\frac{1}{2}$ inch thick. These were steel reinforced belts that were very strong.. The belt passed over a drum at the top of the building, and another below the bottom floor. The drums were powered, and ran continuously. Along the full length of the belt were standing steps on seven-foot centers. Between the steps were $\frac{1}{2}$ inch round bars bent in a 'U' to be used as handholds. Both the steps and the handholds were rigidly fixed to the belt.

The belt passed through a hole in each floor about four by six feet. The belt was further housed in its own shaft that was about four foot square. On either side of the shaft and on each floor were openings about three foot wide, and seven feet tall. These were entry points for a person to step on or off of the belt. The belt on one side of the shaft was going up, and the belt on the other was the return and going down.

A potential rider stood in front of the opening, readied himself, and then stepped into the shaft and on to the moving step. He grasped the handhold, and was moved upward or downward depending on which side of the shaft he had entered. When arriving at the floor he wanted to get off, he timed his step off to just the right moment to allow a smooth transit to the floor.

This all sounds a bit risky compared with our modern day elevators. However it was very efficient, and made for minimum transit time between floors. The first few times I rode the belt were a bit scary, but I soon was able to confidently step on or off of the belt with ease. I did not witness any accidents with the belt. However, it was easy to imagine what happened if a foot slipped off of the step, or if a handhold was let go.

1962 was about ten years before the Occupational Safety and Health Act, (OSHA) was signed into law by the U S Congress. After OSHA, devices like the Man Lift were shut down, and never permitted again. Further, OSHA required that any machine to be operated by a human be made idiot proof. Thus, no idiot could put any part of his body in harms way. Many protective shields, and hand 'pull backs' were installed. All in all, the work places were made much safer, and bodily injuries were significantly decreased. Those of us in charge of manufacturing plants did not always agree with the OSHA inspectors. I found that being cooperative with them made for a smooth transition to a safer work place.

At Hyser Electric, there were many other big and small design projects. Another notable one was a packaging line for a small powdered milk drying plant in Ogdensburg New York. Once again I went to the plant to facilitate installation. I was there for two weeks at first, and then back again after a short home visit. Work was going slow through no fault of mine, plumbing and electrical contractors were delayed. I was anxious to be back home every night with my new bride. The plant Manager was sympathetic with my desire, and facilitated (paid for), bringing Cathy to Ogdensburg. We rented a small apartment in a motel that was directly on the St Lawrence Seaway. It was summer time and there was a big lawn leading down to the water. We were able to do our own cooking or go out to some very nice and reasonably priced restaurants. We could watch the big freighter boats transit the waterway right from our front yard. All in all it was very pleasant. The only problem was that I was working about 60 hours a week. The project finished up about six weeks later. The plant Manager was happy to have his line running smoothly, and we were happy to return to our home. Later in 1987, LaVonne and I returned to this very same town aboard the Vonnie T. We tied up overnight at the Ogdensburg city pier. All are very good memories.

Hyser/Thiele had some standard machines that could be used in a variety of situations. They decided to collaborate with Stainless and Steel Products Co. 'Stainless' built trailer, and truck mounted tanks of stainless steel for hauling milk and a wide variety of liquid chemicals. It was a good match because of the expertise in stainless steel that 'Stainless' brought to the table. I was tasked with helping the transitioning of production to 'Stainless'. During that time I became familiar with the new company's management, and was offered a job with them as Product Development Engineer. It was a much bigger company. There were about 100 employees, 70 shop people, and an office staff of accounting, engineering, and sales. We had six field salesmen all over the United States. The Manufacturing plant was pretty big, about one half of a city block in area.

At that time Cathy and I lived in Columbia Heights MN in a rented duplex. We had two beautiful daughters, Colleen and Mary Claire.

I remember the date of November 22nd 1963. I went home for lunch, and on the way back to work via St Anthony Blvd, and in front of Francis A Gross golf course the news came over the radio. John F Kennedy had been assassinated. Most every one who lived through that remembers where they were and what they were doing at that infamous hour.

Smoking

I started smoking at about the age of 14. It was cool, all of the cool people did it. I became addicted to it. In those days none of the health problems associated with smoking were known. Early on, I could buy a pack of Viceroy's for 17 cents. In my mid twenties, I was smoking two packs a day. While being addicted, I knew I could quit because I would give up smoking for lent some years, real sacrifice. One Friday evening while living in that Columbia Heights apartment, I felt mild pain in my chest when I would take a deep breath. I quit then and there cold turkey. I remember the pangs of withdrawal. The first few weeks found my hand often searching in my breast pocket for the pack of cigarettes that was no longer there. I remember trying to get close to people who were smoking and get a bit of that delicious second hand smoke. In restaurants, every table had an ash tray, and people would often light up right after a meal (that's when the cigarette tasted the best). I tried to get close to those smokers. It took about a year for the fiendish addiction to go away. I remember knowing when it was over. I would now try to get away from smokers. Yes, the second hand smoke was objectionable and obnoxious.

Experimental Stress Analysis

My early work at Stainless was improving the structural integrity of the over the road trailer tankers. We employed a professor from the University of Minnesota Aeronautical Engineering Department by the name of Alan Blatherwick. He had a PHD in structures. I was assigned to work with him.

The plant had a repair department that fixed tankers that had been in accidents, needed warrantee repair, or needed refurbishment because of old age. In that repair department we were able to see some of the design problems that caused cracks in the structure. The first such problem that I was assigned to was the redesign of 'side channels' that held up the tank.

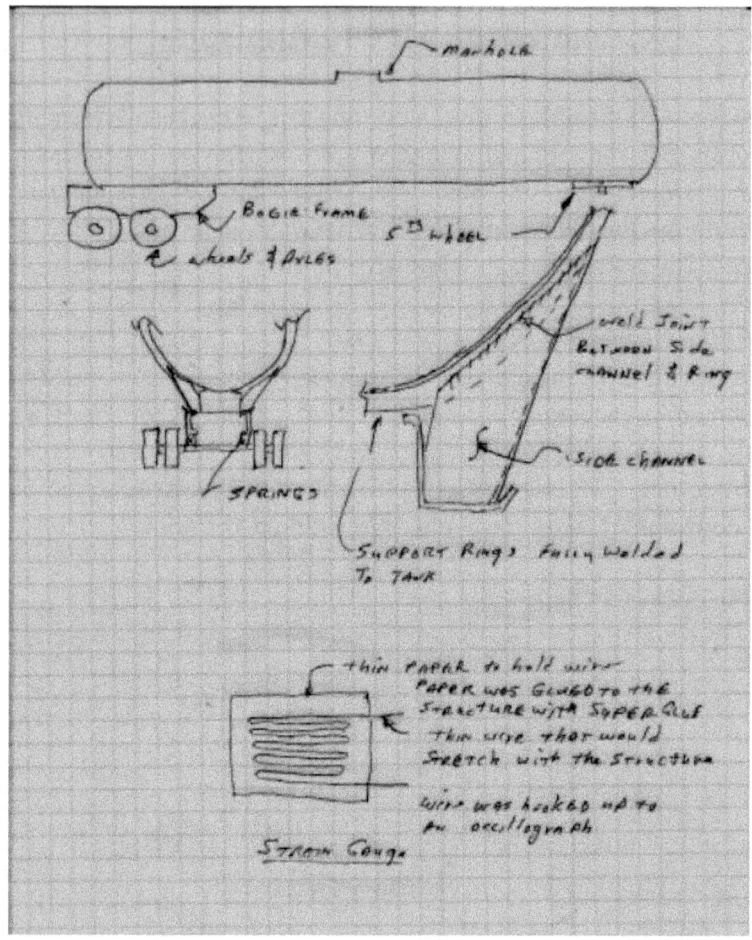

The failures had happened near the top of the side channel where the load was delivered most intensely. Those failures happened only after many repeated cycles of loading. The failures were a result of the stress in the structure exceeding fatigue limit for the material. We tried a few differently designed side channels. To test each design, we would carefully clean several spots near to where failures had been observed and glue on strain gauges with super glue.

Let me digress. There are two kinds of stress analysis, Theoretical (done mathematically), and Experimental (done at the site of suspected problems with strain gauges, photo elastic analysis, or loading to failure). Metallic structures that are loaded in various ways actually stretch under load. That stretching is called strain, and it is measured in inches per inch. If a one inch piece of metal grows to 1.0005 inches under load then it had a strain of .0005 inches per inch. Stress in the metal is what actually causes failure. Stress is the load divided by the cross sectional area of the specimen. Thus, if we have a rod that is one square inch in cross section, and we put a stretching load on it of 50,000 pounds than it feels a tensile stress of 50,000 pounds per square inch. If we are in a laboratory, and can measure the load, and the cross sectional area then all is fine. However if the structure is complex, and cannot be examined in the lab, there is an alternative. If we can measure the strain in the structure, then we can predict the stress by applying the 'Modulus of Elasticity'. There are tables that give us the modulus of elasticity for various metals. The Modulus of elasticity is found in the laboratory for each metal. The stress and strain for a material under load are observed by stretching, compression or twisting the sample. The length of the sample is measured before and after a load is applies, and the 'strain' is calculated. Likewise the load that is put on the sample is measured, and the cross sectional area of the sample is measured. The 'stress' is then calculated.

Now, if we divide the stress by the strain the result is the **modulus of elasticity** for that material. That number never changes as long as the composition of the material is constant.

So if we know the strain in a material we can calculate the stress. Experimental tables have been made for many materials. Those tables contain ultimate (rupture) strength, yield point (stress at which the material cannot elastically return to its original size), and the materials Modulus of elasticity.

High grade steel has an ultimate strength of 70,000 psi, a yield point of 50,000 psi, and a modulus of elasticity of 30,000,000 in tension.

If the metal is under repeated loadings, we must not let the stress get to the yield point. That is also called the fatigue point. We design to stresses that are well below the yield point. In our work with side channel redesign we were looking for a safety factor of about two. Thus, if the maximum stress we observed was 25,000, we had a safety factor of two.

OK, now that the Mechanics of Materials primer is over, let's get back to the shop.

The strain gauge is carefully positioned, and glued to the structure. There may be as many as 50 strain gauges glued to our over the road tank trailer structure. The thin wire in the strain gauge stretches right along with the metal when it is under load. The electrical resistance of wire is related to the diameter of the wire. As the wire in the strain gauge is stretched, its diameter is reduced. If we can measure the change in electrical resistance of the wire, we can calculate the reduction in the wires diameter, and the strain in the material.

A wide variety of strain gauges are available depending where they are meant to be used. Each of those strain gauges has been calibrated so we can get strain directly from measuring the change of electrical resistance.

Measuring, and recording changes in electrical resistance is no small problem. We hooked up the strain gauge into a Wheatstone bridge circuit. We could null out the voltage in the bridge by using a variable resistor in one of its legs. Under no strain, the voltage was zero. However when we applied a load to the structure, the metal strained, strain gauge stretched, the resistance of the wire changed, and we could measure that change in the Wheatstone bridge circuit manifested by voltage change.

We used a recording oscillograph to track what happened with each of about ten strain gauges hooked up to it.

Now, let's go on the road. The trailer tank was coupled with a tractor, and filled with 5700 gallons water. We found an isolated roadway on which to run our tests.

The oscillograph was placed in the sleeper cab of the tractor, on a nice soft mattress. Doctor Blatherwick sat beside it and directed things.

As the trailer passed by a certain point, two of us threw 4 x 6 timbers in front of the trailer wheels, and behind the tractor wheels. In that way, the tractor with its sensitive instruments was not bounced about, but the trailer was made to feel a simulated large chuck hole in the roadway. The oscillograph recorded the strain on the structure at the moment of impact.

We repeated this test over and over again, and at different speeds until we believed we had a good set of data.

Later, we analyzed that data and determined how much stress was present at the site of each strain gauge during maximum loading.

Repeats of the testing with various designs of side channels resulted in selection of a design that minimized stress under high loading. The top of the chosen side channel met the ring tangentially. At that point the face of the side channel tapered into a tongue about 4 inches long. That tongue was fully welded to the ring.

The new design was successful, and as far as I know that design is still in use.

Similar testing was done on other parts of the trailer tank structure.

Another stress analysis tool that I liked was Brittle Coating. Those coatings are similar to varnish, but sprayed out of a can. The brittle coating is relatively inelastic. When sprayed over an area of interest in a complex structure, it dries, and firmly bonds to the structure. Then when load is applied to the structure, the metal stretches, but the coating does not. The result is small cracks in the coating. Those cracks first occur in areas of high strain. Those cracks are exactly perpendicular to the strain direction. Also, the distance separating the cracks is proportional to the strain in the metal. I used brittle coatings to locate strain gauges that could accurately measure strain.

Organizations

I joined the local chapter of SESA (Society for Experimental Stress Analysis). We met monthly. The members were engineers from local companies that built complex structures. I liked it so much that I volunteered to become Vice President. My main task was to line up speakers and/or trips into the various shops that made the structures.

Interestingly, about that same time, I explored membership in the Junior Chamber of Commerce. Somehow, since I had lived in Minneapolis, and worked in St Paul, there was a choice to be made about which organization I would join. Well, someone thought that might make a good story. They arranged to have me stand in the middle of the Franklin Avenue Bridge that separated Minneapolis and Saint Paul. Then they brought Miss Minneapolis and Miss Saint Paul to the middle of the bridge. Each of these very beautiful young ladies grabbed one of my hands, and tried to pull me toward their city. A photographer was there to record the event, and the 'tug of war' story showed up the next day in the Minneapolis Tribune, and in the Saint Paul Pioneer Press. What fun!

First House

We saved and scraped enough money up to buy our first home. It was on the Corner of Bedford Street and Franklin Avenue. It was a two story duplex with three bedrooms on each floor. It was in bad shape, but structurally sound. It was about 75 years old. The down payment was $1000, and the bank loan was $18,000. We first lived in the upper unit, and rented out the unit on the main floor. I was very ambitious in those days, and was determined to bring it up to good condition, and raise the rents. The value of rental property is about ten times the net operating income. Thus, if I was able to get a 20% increase in the rents, the net operating income might go up by 30%. I was able to do that over a period of a couple of years, and add about $10,000 to the market value of the building.

However it was a lot of work. I removed four layers of wallpaper from the walls and ceilings, painted all of the woodwork, and sanded and varnished the fine maple floors. I re stashed the stucco exterior, and repainted the trim. The cabinetry was updated and new sinks were installed.

The duplex had a finished bedroom on the third floor and in the attic. We painted that room and furnished it. Brother Mike came to live with us in that room during his first year at the University. It was a long haul up three flights of steps to get there, but it had a smashing view of the wooded street in front of us. I was glad to have Mike there.

The heating system was an oil-fired boiler in the basement that heated circulating water. A pump pushed the water up to the 1st and 2nd floors and into radiators. A thermostat in each unit controlled how much hot water would be delivered.

At the time I had been experimenting at work with epoxy adhesives and putty. Lucky for me!

On one cold winter day, we discovered that we had no heat. An inspection trip to the basement exposed the problem. One of the bolting cast iron 'ears' that was used to join the upper and lower halves of the furnace had a crack in it. I devised a plan to fix it.

First thing was to hook up a hose to the boiler so the leaked water could be replaced. Then after the system was full of water again, I turned on the heat. Yes the furnace still leaked, but the hose kept the system filled. I let it run like this until the temperature in the units was about 80 degrees. Then, I shut the system down, and drained it. Next was to grind down the area of the crack so as to make it clean and dry and ready for my epoxy putty patch. I did not know if this would work, but we couldn't afford a new furnace. I mixed the putty, and pushed it into and over the crack. To assist with the curing process, I put a reflectorized light bulb near the repair area. The light raised the temperature of the putty, and decreased the 'set' time. A few hours passed before I thought the putty to be cured. Coincidentally, the temperature in the living units had only dropped to about 60 degrees.

I filled the system with water, and vented all of the air out of each radiator.

VOILA! It worked, and continued to work for at least a few more years. When we sold the duplex, the patch was still working.

Those were good years. One of our tenants was Jewish, and became good friends of ours. We were honored to be invited to the 'Circumcision rite' for their first-born son. The Rabbi came into the apartment and did his thing, and a gala party followed.

I also remember my first daughter Colleen at about the age of five falling on the cement steps leading up to the duplex. The cement was old and had a small chunk broken out of the front of the step. Colleen fell on that broken cement, and cut herself at the bridge of her nose. We didn't take her to the doctor for stitches. In those days, doctors were not used as much as they are now. The result of the injury is a small scar that is probably not noticeable to most. I noticed it, and felt guilty for not having patched that step.

Olson Vent

OK, back to the shop. They say necessity is the mother of invention. Over the road milk tankers had three problems with ventilation.

Surge slop was caused when a partially filled tanker (say 2/3rds) was required to come to an abrupt stop. The milk would surge forward against the front head, and top of the tanker. At the manhole, it would erupt outside of the vessel if not contained.

Secondly, while pumping milk into the tank from the rear valve, it was possible to build up a high pressure. If the vent was inadequate to relieve the air that was being displaced by the incoming milk, damage would result from high pressure.

Third, while unloading milk through the rear valve, damage could occur if the manhole vent would not let in enough air to satisfy the vacuum that was caused by the departing milk.

There was nothing on the market that would solve all of these problems.

After a lot of thought, and experimentation with various devices, I came up with what is now known as the 'Olson Vent'. It is still the standard for the industry after 55 years.

We had a repair shop that welcomed any sort of a problem with stainless steel tanks. When a large over the road milk tanker had been unloaded without ventilation the result was a 'suck in'. The driver was supposed to open the manhole cover prior to dispensing his load. If this was not done vacuum formed and pulled the top of the tank downward and inward. The point at which the damage starts was about 5 psi in round cylindrical tanks and about $\frac{3}{4}$ psi in oval tanks. Similarly, if the tanker was loaded from the rear valve with no ventilation, the pressure would make the heads of the tank move outwardly as they tried to get spherical. The result was severe buckling near the seam between the head, and the tank shell. The point at which damage occurs from pressure is about 7 psi in round tanks, and about 2 psi in oval tanks.

How did I know this? I used old tanks that were once in service, but were lying around the grounds waiting to be sold for cheap storage. I set up experiments where I would pump water in or out of the vessel with a variety of vents that I had been working on. I would observe the pressure inside the tank as the filling or draining proceeded. I could tell from the sounds, and shape of the tank when the damage point was close.

As a matter of fact, while doing the design/prototype work, I accidentally partially exploded an old tank. An early design vent was installed in the manhole cover, and I pumped water into the tank much like a milk hauler would pump out milk from the farm tank and into his pickup tank. I wasn't careful to stop the pump when the tank was full. My vent would not pass huge amounts of water like it did air. I over pressured the vessel.

I was a little scared. The tank had some value, maybe a thousand dollars before my experiment, and zero after. I was worried that my bosses wouldn't like that too much. However, and a big plus, I now had empirical data to show at what pressure the damage would occur. I could then duplicate the test setup in a heavy pipe without worry about damage and my experimentation went on.

I took photos of the inside of that partially collapsed tank, and we were later able to use those in advertising layouts for the vent.... "Here is what can happen to your tank if you pump in without having a working vent on top to satisfy the pressure created". All milk haulers knew about the probability of such damage, and were always on guard to have the manholes open during filling of draining. My new vent assembly would minimize the potential for damage if the manhole was not opened.

As an aside, I learned an important lesson about electrical shock. When I pumped water into or out of an experimental tank, it was always on the repair shop floor. That floor was always wet. A 220 Volt motor powered the pump that pushed water into the experimental tank. That motor was hooked up the shop electricity through heavy-duty cords. Those chords were joined together by heavy-duty metal cased plugs.

OK, you can probably see this coming.... One day when I was hooking up the power, I grasped the two plugs to twist them together. A huge amount of 220 power went through the plugs, and into my body. I was thrown about 6 feet across the floor, and landed in a heap. No serious damage was done. Later, in analyzing this event, I discovered that a strand of the braided wire inside of the plug housing had been carelessly allowed to contact the metal jacket of the plug. The maintenance people who had installed the plug made a mistake.

From that day forward, I never could touch one of those plugs.

As another aside, I had a little of practical joker inside of me. One day I caught a small mouse on the shop floor. He was a cute little guy, and very wriggly. Now it turns out that the receptionist/greeter person was also a bit of a joker. She had pulled a few practical jokes in her time. She had a bubbly personality and we all liked her. My devilish mind conjured up a plan. I put the little mouse into my pocket and went into the office. I waited for a time when the receptionist was away from her desk, and there were no other people in the area. I then opened her top desk drawer, and put the little mouse inside before closing it.

Now guess what happened the next time the receptionist opened the drawer. The poor little mouse was very scared, and leapt out of the drawer with maximum vigor. The receptionist let out a very large "EEEEEEEK'.

I was careful not to tell anyone about my prank.

One more aside. I was on a business trip to New York City. We went into the Metropole Bar and Restaurant on Times Square. We sat in a booth and were entertained by Duke Ellington. It was a very nice place. I sat on the aisle, and had my size 14 shoe out a bit into the aisle. A man walked down the aisle and tripped over my foot. I got up to apologize, and realized it was Cassius Clay. You know him now as Mohammed Ali. Cassius accepted my apology, and we talked for a minute or so. He autographed my menu for me. At that time he was the best amateur boxer in the nation. He turned professional shortly thereafter, and won a lot of bouts. I kept that autographed menu for some time, and then lost it. What would that autograph be worth today?

The double acting vent was made to sit inside the ferrule mounted on the manhole cover. It comprised of a body with a pressure opening, and vacuum openings, a pressure plug, a vacuum plug, and a spring to tie the two plugs together. The spring exerted the same force on each plug, but because the pressure opening had a relatively small area, the pressure plug would open at a higher pressure. The vacuum plug was situated on the lower side of the body, and covered three annular slots. When milk was being pumped out, and the resulting vacuum increased, the vacuum plug would open and admit air to satisfy the vacuum. Here is a photo of the assembly, and the parts.

T 21642 Vent Assembly

The beauty of the device was its simplicity. It was two relief valves built into one assembly. Again, as the pressure in the tank increases, the pressure plug would move upward to release the air pressure inside the tank. As the pressure in the tank would decrease due to unloading, the vacuum plug would move downward, admitting air to satisfy the vacuum.

The whole device had to pass sanitary standards. The materials, Nylon, Stainless Steel, and Silicon Rubber had all been approved for use in milk service. Additionally, all of the parts had to be easily cleaned so that no bacteria would build up and potentially contaminate a load. I designed each part with a minimum inside radius of 1/8th inch.

Two models of the vent were built. One was for round tanks, and one for oval tanks. Each type of tank required different opening pressures.

The additional benefit for this vent assembly was that it was able to contain all but the most violent surge loadings that resulted from rapid stops.

Along the way, I filed 'Record of Invention' documents. Later I began the patenting process with a search of the patent office in Washington DC. That search was done by a firm specializing in such searches. The result of the search was that there was a similar device already patented for use in gasoline tanks.

Let me digress. In those days, there were two types of patents. A letters patent was used to disclose new art, or science. Thus, no one could copy the original idea even if parts of the invention were varied. A picture patent was used to patent the exact device. It had little value since one could copy the invention, change one component, and it would not infringe.

We decided not to pursue a patent for the Olson Vent T 21642. The number was assigned by me to fit in the archives of drawings at Stainless and Steel Products. The 'T' indicated it was a tank produce, the 1 indicated the drawing was made on 8 ½ by 11 drawing vellum. The other numbers made it unique to my drawing.

It turned out that a Patent was not really needed. We tooled up for production buying molds and spring making equipment. That equipment had a cost of about $20,000. Thus, anyone that wanted to copy my design would have to pony up the money for it.

We put the vents into production and installed them on all of our milk tankers. The reputation of the vent was soon established by the people who hauled milk. I knew we had a winner when our competitor tank manufactures started buying the Olson Vent from us.

The vent is still the standard for the industry, and can even be bought 'on line'

http://www.tankerpartsstore.com/mm5/merchant.mvc?Screen=PROD&Product_Code=OLS-T21642&Category_Code=SANVENTS

Urethane Foam

Prior to my being employed at Stainless and Steel Products, The Company used Urethane foam as an insulant in the space between the inner head, and the jacket head on milk tankers. Again, a milk tanker is comprised of an inner shell (which is also a structural member) and an outer shell. The outer shell serves to confine about 2 $\frac{1}{2}$ inches of insulation, and make a clean looking exterior on the tanker.

The Urethane form was mixed in a machine that combined Polyol resin and Prepolymer with Freon 11. The Polyol resin and the Prepolymer combined chemically in an exothermic reaction (the reaction gave off heat). As the temperature of the mixture rose, the Freon component began to boil, having a boiling point of about 75 degrees Fahrenheit. The boiling took place in very small cells caused by the thorough mixing. So, as the reaction started, the Freon boiled and turned to a gas. That caused the mixture to expand. If one would shoot some of the liquid mixture into a box, it would slowly expand as the Freon boiled from the heat generated. The box would slowly fill with warm (about 125 degrees) foam. As time went on, the chemical reaction would be completed, and the urethane would solidify. After solidification, the foam would gradually cool to room temperature, and slightly shrink in volume.

Urethane foam has about twice the insulating value than fiberglass. Urethane foam in its liquid phase is very sticky. It the substrate (steel in our case) was coated with a suitable primer such as Zinc Chromate, a tenacious bond would form between the foam and the steel.

We wanted to use Urethane foam as a structural insulant in the total annular space between the inner shell and the jacket of milk tankers. However we found that when we poured the foam mixture directly from the mixing machine and into the annular space a problem arose. As the foam rose (expanded) in the annular space and filled it, the foam would set up or harden. After setting up, the foam would shrink as it cooled to room temperature. That shrinkage, when looked at in cross section, was disastrous for the jacket of the tanker. The inner shell, being of relatively heavy steel, and cylindrical in shape would not move. So as the shrinkage occurred, the foam tried to pull the jacket material inward. The result was that the foam pulled away from some parts of the jacket and not from others. Now we had wrinkles all over the jacket of the tanker, and the bond between the jacket and the foam was essentially destroyed.

United States Patent #3303617

We wanted to use Urethane foam as a structural insulant in the annular space of over the road tankers.

To digress, the tanker can be thought of a forty foot beam continuously loaded with it's liquid payload, and supported at the back end through the bogie frame and thence through the springs, and axles, and tires to the ground. It is supported at the front end through a trailer kingpin and 'fifth wheel plate', and thence to the frame of the tractor that would pull the trailer. That forty foot beam under load would be subjected to compressive stress in the top, and tensile stress in the top. The compressive stress in the top can cause buckling if the structure is not strong enough.

Normal construction at that time called for 12 gauge stainless steel to be used in the top of the tank. We wanted the use the structural insulant Urethane Foam to bond the inner and outer shells together, and make them both be load bearing members. If that could be done, we might be able to reduce the thickness of the inner shell to say, 16 gauge. This would save a lot of money in the reduced cost of the shell material. The corresponding weight reduction could be used to make the tank bigger, and thus carry a bigger payload. But alas, the foam shrinkage would not allow that.

I set out to devise a system that would overcome the bad effects of Urethane Foam shrinkage in a structural sandwich.

I had a fixture made that would let me put a 16 gauge 12 x 36 inch piece of steel on one side of the fixture, and opposite it on the other side of the fixture and 2 $\frac{1}{2}$ inches away, a piece of 20 gauge steel. I sealed the ends and bottom of the fixture, primed the steel with Zinc Chromate. I then filled the cavity with Urethane Foam. I observed what happened as the foam started to cool. With a dial indicator, I could measure the shrinkage. And of course it shrank about 1/8th inch.

I thought and thought, and came up with a plan to 'precompress' a portion of the foam, and let that precompression compensate for the later shrinkage of the foam that was to complete the sandwich. Thus, the portion of foam that was solidified, would compress (precompress) with the pressure exerted by the rise of the secondary foam. That precompression would be allowed to decompress and expand as the secondary foam fill shrank and contracted while cooling.

Now I needed a spray gun foam mixer to apply the precompression coat to the 20 gauge part of the sandwich. I sprayed the 20 gauge piece, and then machined down the solidified foam so it was a uniform ½ inch in thickness. I put the pre foamed 20 gauge steel into my fixture, and placed the 16 gauge piece opposite it. The resulting open cavity was 2 inches wide. I then foamed that cavity, and as the foam cooled, measured the resulting shrinkage or lack thereof. I repeated this test many many times until I was able to hold the original 2 ½ inch total thickness before and after foaming. Therefore, the net effect of the shrinkage was zero. The ideal thickness of the sprayed foam was 1 1/8th inches.

We knew that to use this sandwich as a structural insulant, it would have to stand up to the repeated fatigue loads that were encountered in an over the road transport. I made a lot of those 12 x 36 inch sandwiches, and got Al Blatherwick to agree do a fatigue test on them in the lab at the University of Minnesota. We fashioned a support for each end of the panel. And loaded it with a yoke in the center to a machine that would vibrate it at about 60 cycles per minute. The plan was to get at least 100,000 cycles on the sample, and have it finish the test with no damage to the foam, nor the bond between the foam and the steel.
Many samples later, we deemed the sandwich worthy of functioning properly in the fatigue environment.

OK, now how the heck do we duplicate the spray coating process on large pieces of steel?

I thought and thought, and remembered my experiences with floor sanders used on Maple floors. I bought a floor sander and equipped it a coarse grit sanding belt.

Next was to lay out the actual jacket material for an experimental full size model of an over the road tanker. For a trailer tank, these pieces would be 4 feet wide, and about 17 feet long. I laid out the steel on the floor of my little in shop lab, primed it and sprayed it with more than 1 1/8 inches of foam. Next I built a frame for the belt sander to ride on. At each end of the frame, I mounted ball bearing wheels of a VEE shape. They would ride on an angle iron track that was fixed to the floor on each side of the spray foamed jacket piece. I adjusted the height of the sander assembly to be 1 1/8 inches off of the floor. I started up the sander and trimmed the foam to the desired thickness.

After the jacket sheets were finished we installed them on the tanker. Each foam section fit between the reinforcing rings on the tank. The foam was actually 42 inches wide. A band of solid urethane foam was fitted over each ring to bring the depth out to $2 \frac{1}{2}$ inches. The 3 inch foamless part on the jacket fit exactly onto the ring assembly. Now the whole thing was riveted down at the bottom. After all sections were in place, we foamed the now 1 3/8 inch annular space. When the new foam cooled, there was no bad effects from shrinkage.

The experimental work was done in late 1962. The patent work was done and filed on April 16 1963. The final Patent was issued on February 14th 1967.

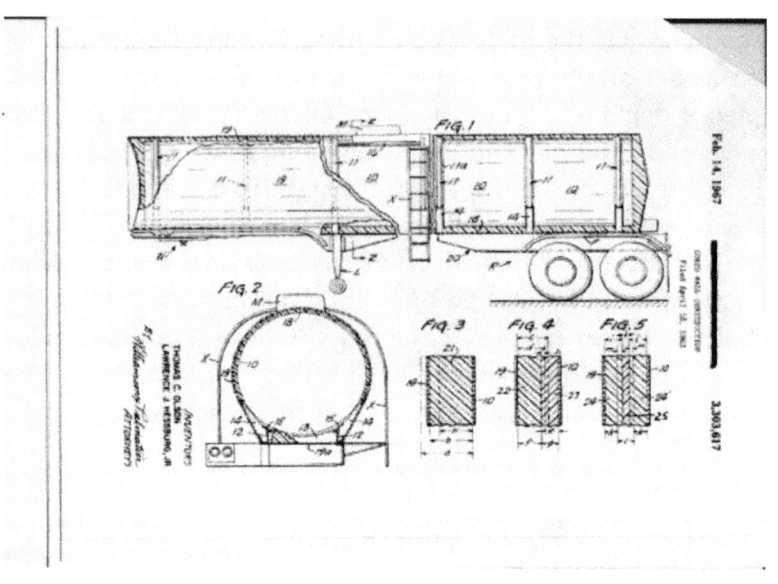

Containerization System

At Stainless and Steel products, we had a milk tanker customer that operated in the mountains of Vermont. He used truck mounted tanks to go to farms in the mountains to pick up milk. The roads in those mountains are narrow, and winding. After each tanker was full, they then drove it about 100 miles to the creamery in Boston where the milk was processed into carton milk, butter, cheese, and a few more things. The customer, Stewart Rouse, thought that if he could transfer two truck mounted tanks to a trailer, he would save money in transportation costs. There would be only one driver, and one rig instead of two on the trek to and from Boston. So Stewart came to us with the idea. We made

62

an agreement with him to design and build a system to implement his idea. He of course paid us for the work, and we gave him the rights to use the system elsewhere.

About this time I got an assistant, actually, a very good draftsman/technician named Steve Zanosky. I also got promoted to be Chief Development Engineer, and was given a raise.

The first problem was to relocate the pumping equipment on the bulk milk pick up tank. I designed an insulated box that would mount on the side of the truck frame (fig2). It had a flexible rubber connection to a valve assembly on the tank.

The second problem was how to mount the tank to the truck frame in such a way that it could be quickly unlatched from the frame, and provide it with the means to be lifted up and onto a semi trailer frame. The locking mechanism is shown in figure 4. I designed a lifting beam that was welded to each end of the tank frame. That beam was a large heavy gauge rectangular tube that extended out to near the full with of the unit. Over the rectangular, I placed another larger rectangular tube that would telescope over the beam (fig 5). Thus, when extended the telescoped tube would come out from the sides of the unit, and provide a point from which to lift the tank upwards for transfer to the trailer.

Now, we built a semi trailer that had mating locking devices for the tank (fig 3). The next thing was to build a transfer station that would remove the tank from the truck, and hold it suspended until a second tank arrived. Once the second tank arrived, the semi trailer was backed under the two of them. The two tanks were lowered onto the trailer, and locked in place. The trailer would then make the trip to Boston and back, and the whole process would repeat.

The transfer station (fig 6) was something new for me. It was driven by hydraulic cylinders, and powered by a large hydraulic pump. Most of the hydraulic design was given to me by the vendor of the hydraulic mechanisms. Sizing of all of the components required careful analysis. Each of those tanks weighed about 20,000 pounds.

When all of the hardware had been completed, and moved to the transfer site, I went along to supervise construction. The site was in Bellows Falls Vermont, and on the Deleware River that separated Deleware from New Hampshire. The construction took two or three weeks, and I went home for weekends. I learned a little about local customs there. The people spoke with a mountain drawl that was sometimes hard to understand. When you asked for coffee, they asked if you wanted it 'regular'. That meant that it was very strong, and laced with cream and sugar. I liked those people.

United States Patent
3421646

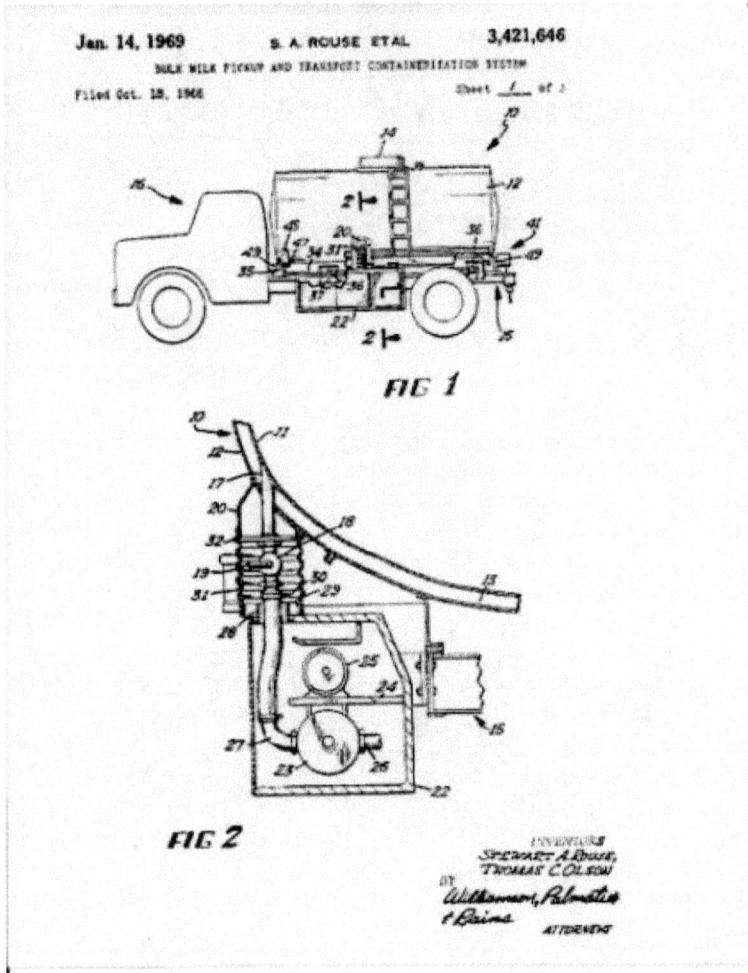

FIG 1

FIG 2

INVENTORS
STEWART A. ROUSE,
THOMAS C. OLSON
BY
Williamson, Palmatier
& Bains
ATTORNEYS

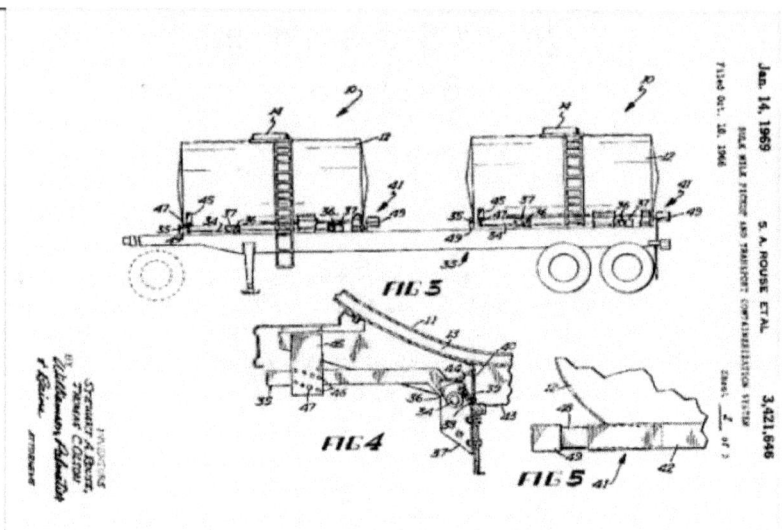

FIG 3

FIG 4

FIG 5

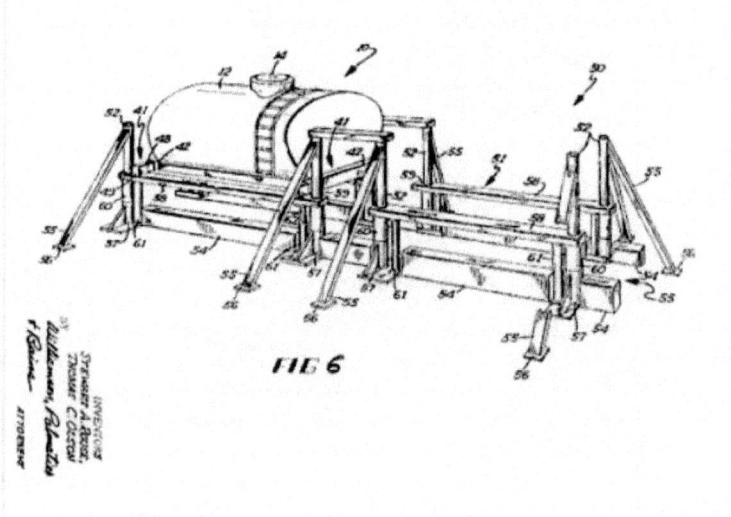

FIG 6

66

Dear Reader

I am glad you made it this far. I know that a lot of this stuff is dry, and a lot is pretty complicated. I am trying to write it as simply as possible, and yet transfer the gist of each mechanical adventure.

I am writing this so that you may know a little about me. Of course it isn't possible to write about the whole me, and I have left out a few things I choose not to relate.

I had three grandfathers; George Cloutier, Carl Olson, and Ba (a name I gave to my maternal grandmother's second husband). I know very little about any of them. They each no doubt had a wonderful story of their lives that was never written down. I regret not knowing each of them better, and not having the Memoir that each may have written.

The same of course, is true for my grandmothers Ellen Dorn, and Hanora Cloutier.

So now, I encourage you to write something about yourself so that the generations that follow you may know something about you. I am very sure that many of you will rise to do very kind things, and very great things. Write them down!

I am trying hard to sprinkle this writing with some of the family glimpses that make all of our lives so rich.

House Number Two

After all the fix up had been done on the Bedford Street duplex, we moved to a single family house at 2194 Dudley Ave, St Paul. The house was only about three blocks from the University of Minnesota agriculture campus.

 It was a two story house with three bedrooms. It was about 70 years old. It had a nice little goldfish pond in the back yard. We foresaw a problem. With little kids bomping around the place we were afraid that one might fall into the pond and drown. That was not to be. We filled in that nice pond with dirt, and transformed it into a flower garden.

The house was in pretty good shape, but it had a small kitchen and adjoining back porch overlooking the back yard. I removed the wall between the kitchen, and porch, and finished the new large kitchen with fresh flooring, and cabinetry.

Andy and Sara were born to us on Dudley Avenue. Life was good.

When Mary Claire was about 6 years old she exhibited an unusual talent. She was left handed. When she wrote a sentence on a piece of paper, it came out fine if she started on the left hand margin. If she started the sentence on the right hand margin, the sentence would come out upside down, and backwards. If you held up the paper to a mirror, it became perfectly legible. I thought, how cool. She might be able to see both sides of any situation she would be in. Some time later, a Doctor told us to discourage her writing mirror image for fear that it would affect her brain. I didn't understand that, but we complied. Today at age 48, she cannot do the mirror image thing. I learned last week that Leonardo DaVinci also had the ability to write mirror image. In fact he wrote all of his diaries that way so others would not be able to read them.

I remember buying a brand new 1963 Chevy Belaire car. It was a couple of months before Christmas, and we were gathering Christmas gifts for the kids. We carefully wrapped each gift, and placed it in the trunk of that car. About a week before Christmas, we awoke to find that the car had been stolen. We thought about our loss, and especially about those missing presents. Luckily the police recovered the car the next day. Apparently, someone punched out the ignition, and went joy riding with the car. The trunk had not been opened. How lucky was that?

While living on Dudley, we rented out each floor of the duplex on Bedford Street. I remember getting occasional phone calls from the tenants about toilet problems and the like. The calls were not very frequent, but it was getting so that every time the phone rang, I thought "oh, what now". I looked for an opportunity to trade up. As it turned out I found a guy that owned a 14 unit building in south Minneapolis. He needed cash. He sold us a half interest in the building. Our income from real estate took a nice little jump. Managing that building was not always easy but now I had a caretaker to act as a buffer between the tenants and me.

Boat Number Five

While riding past the boatyard on Smiths bay one day, I noticed a sign that advertised a free trial ride in a sailboat. Well, that sounded like a good idea. I took the ride with the salesman. The boat would go down wind (broad reach), sideways to the wind (beam reach), and somewhat upwind at about 45 degrees to the wind (a beat). I thought it was fun, the darn thing would do all that without an engine simply by harnessing the wind. It could return to the same spot it started from. This boat was pretty small, about 12 feet long. It could carry only two people, and as it was new, it was pricey.

I kept that in mind, and was soon to find a 19 foot class Y boat, a sloop made by the White Bear Lake Boat Works. I bought that boat because it fit my budget, and was big enough to take four or five passengers. However, it needed some work.

I trailered that boat over to the East end of West Arm, and launched it alone. This would be the second time I had ever been in a sail boat, and I was not sure I could sail it let alone getting it back to the same place I started from. Well, it was a nice day with a mild 8 to 10 knot wind blowing. I pushed off, and raised the sails. Lo and behold, I could sail at a pretty good speed (maybe 6 knots). I played with that boat for an hour or more, and came to know about beating, reaching, and sailing off of the wind. I was very impressed by what that boat could do, and pleased with myself for catching on to it so quickly.

I took that boat to the folks place at Casualaire, and fashioned a mooring for it using a couple of cement blocks for an anchor, and a big foam float from which to fasten a mooring line. I went back home to St Paul that day with fond memories of my boat bobbing at it's mooring. We had a lot of fun with that boat. My love affair with power boats had ended.

Y Boat

Cryogenics

We were always looking for challenging jobs to make money for the company, and to learn something new. We bid on an offering from NASA (National Aeronautics and Space Administration) that wanted a 500 gallon tank for transporting liquid nitrogen. The tank was to be used to haul liquid nitrogen that was later transferred to smaller vessels. These smaller vessels would then store biological specimens to be used in the space program. Well, we were the lowest bid, and we got a contract to build the little tank.

We were not totally unprepared since we had developed a new relationship with a company in New Prague Minnesota called Minnesota Valley Breeders Association. MVBA collected sperm from bulls, and froze it in small vials in a liquid nitrogen filled tank. This sperm was gathered from live bulls in a very daring manner that I will leave to your imagination. The semen collected could be stored in the liquid nitrogen for many years before being used to impregnate female cattle by artificial insemination methods. The idea was to use some of the semen immediately on a variety of cattle, and then trace the resulting offspring for high quality milk. After the semen had been used to sire high quality cattle it could then be sold to farmers at a premium price. The process often took five to ten years, but the frozen semen did not deteriorate with age.

MVBA made cryogenic tanks to be used to transport the vials of semen. They sold these small tanks nationwide.

Liquid nitrogen boils at minus 320 degrees Fahrenheit. That is pretty cold. Bear in mind that absolute zero is minus 459 degrees F. Absolute zero is the lowest temperature that can be achieved. It is the point at which the entropy and enthalpy of a cooled gas reached the minimum. If you pour liquid nitrogen into a container at room temperature and pressure, it rapidly evaporates.

So the challenge was to design a vessel that minimized heat transfer from the atmosphere to the liquid nitrogen.

Heat is transferred by three methods:

Conduction is the flow of heat through a substance. Say you put one end of a steel rod into a fire, and try to hold onto the other end. If the rod is short the heat will be quickly transferred, and your hand will burn. However if the rod is long, it will have more resistance to conductive heat transfer. If the rod is a slender tube that has a minimum cross section, the extra conductive heat resistance of the thinner metal will further minimize the heat transfer. So, to minimize conductive heat transfer, you use a rod of minimum cross section, and maximum length. Some materials are very conductive such as copper, and some are not conductive such as wood and plastics.

Convection is the transfer of heat using a fluid medium. If the fluid medium is hot on the bottom, it becomes lighter, and rises through the cooler medium. The cooler medium then falls because of its higher density. If a gas such as air is between a hot and a cold surface, molecules of air will touch the hot surface, and take on heat. Then when they randomly touch the cold surface, they will give off that heat. If there is no air, or other fluids between the two surfaces, there will be no convective heat transfer.

Radiation is the third type of heat transfer. It is also called electromagnetic radiation. This type of heat transfer needs only two bodies of different temperature to work, and is unaffected by insulation between the bodies. The only thing that can change it is the presence of a reflector in the space between the bodies.

The design of our 500 gallon liquid nitrogen tank needed to minimize the effects of all three types of heat transfer from the exterior to our liquefied product.

The tank consisted of an inner shell made of stainless steel, and a heavy outer shell made of high grade steel. The outer shell was mounted onto a frame, and thence onto axles, springs and tires. The inner shell was suspended inside the outer shell.

To minimize conductive heat transfer, I designed hubs to be placed at each end of the vessel. The hubs had three stainless steel rods attached to them 120 degrees apart. The rods were calculated to have the minimum cross section to be able to support the inner shell in the vertical plane. On the outside of the hub was a nylon solid that touched the outer shell, but was not fastened to it. Thus, when thermal expansion took place, it would not stress the outer shell.

To minimize convective flow of heat into the vessel, we evacuated the space between the inner and outer shells. Of course one can never get down to a perfect vacuum, but our goal was to get down to .002 psia. At that pressure, there were very few air molecules present to transfer heat by convection.

Radiation was the toughest problem to solve. The people at MVBA had developed in conjunction with Dupont, a 'Super Insulation'. That super insulation placed a number of highly reflective Mylar sheets separated by fiberglass into the space between the inner and outer shells. The principle was that the electromagnetic rays on their way to the cold surface would be reflected backwards. Again, this was not totally efficient, so multiple layers of Mylar and fiberglass were positioned in the interstitial space.

It all worked well. One of the tough things was to make the outer shell impervious to any gas flow through it. Hydrogen gas can slightly penetrate even the best made vessel. The steel was high grade, and has very few flaws. The welding was carefully done in three passes. To test the soundness of the outer shell, we pulled a very high vacuum on the outer shell. We then hooked up a mass spectrometer to the exhaust from the vacuum pump. Thus any leakage from outside in would pass through the vacuum pump, and thence to the mass spectrometer where it would be identified. We used a bottle of compressed hydrogen to flood the areas around the weld, and any other suspected spot. If the mass spectrometer indicated the presence of hydrogen, we noted the location. Later the offending weld was ground out and repaired. The test was repeated until we had no leakage.

Now the fun started. We shipped the little tank to the NASA base at Langley VA for acceptance. I went there to facilitate acceptance. They did no inspection, except for eyeballing the little tank. Everything was OK, and we never heard if there was a later problem. On my trip into the base, I was driven by an Air Force Corporal in a military vehicle. As we drove across the base, I noticed a lot of little garage like structures neatly lined up. There were hundreds of them. I asked the Corporal what they were for. He replied, "Those are for Atomic Bombs". Yikes!

Apollo 11

I had a very small part in man's landing on the moon.

NASA had built the moon rocket over years of development. They determined that before any launch, the liquid fueled rocket engines should be very clean. They wanted to flush them out with Methelene Chloride, Acetone, and distilled water. Their plan was to park a trailer tank filled with the solvent next to the Apollo rocket, pump the solvent through the engine, and into a receiving trailer tank on the other side. They wanted 5000 gallons of each solvent to pass through the rocket engine. This was done on the launch pad just prior to ignition and lift off.

They specified what they wanted each of three sets of tankers to do, and put the job out for competitive bidding. We saw that request for quotation, did some preliminary design and estimating work, and submitted a bid. We won the contract.

Each of the three sets of tankers was different. However, they all had large three inch pumps, and filters. I did some design work, but it mostly required locating vendors to get the required sets of pumps, valves, and filtration equipment. The tanks were all made of stainless steel and ground to a smooth interior finish. That was no problem for us. We had been doing that for years in Milk tanks.

We shipped the six tankers to Cape Kennedy Florida. NASA required that company representatives be there for acceptance and testing. Jack Hessburg, myself, and our two wives made the trip. What tough duty. We went from January in Minnesota to Florida. We rented a small cottage right on the Ocean. Wow. In our off times, we swam in the Ocean, and I even tried surf boarding. We had dinner one evening in a nice restaurant, and were seated at a table next to Walter Cronkite, and David Brinkley. They were there to cover a NASA lift off.

This was, of course, well before 9/11. The security around the base was minimal, especially for invited guests like us. We were given a tour of the base. We entered the Vertical Assembly Building where the Apollo vehicle was in the final stages of assembly. We rode the elevator to the top of the vehicle, and were able to look down into the capsule that was to later take the crew to the moon. The vertical assembly building was said to be the largest building in the world by volume. It was so big that they had trouble with the atmosphere inside. They told us that when first constructed, it would cloud up and rain inside the building. They remedied that with massive air conditioning units.

On launch day, the Apollo vehicle would be removed from the building through massive doors. It was placed on a massive platform with trackson assemblies on each corner. Each trackson assembly had four huge Caterpillar like tracks. Those tracks were about eight feet high and four feet wide. The whole transporter loaded with the Apollo vehicle then crawled along a specially built (and very strong) road en route to the lift off pad.

The whole scenario was very awesome.

It was a very pleasant week. Maybe all those years in engineering classrooms were worthwhile.

Here is a pic of our tanks lined up at Cape Kennedy (later to be renamed Cape Canaveral)

People

My experiences at Stainless and Steel Products were always rewarding, and enriching. I am most thankful for being able to work with all of those fine people. I knew them all pretty well. I had respect for everyone from the Sweeper to the Welders, Grinders, Machine Operators, Foremen, and the Superintendent. They helped me in many ways. My closeness to these people was helpful to me in later years, when as Manufacturing Manager, I was able to lead them to higher levels of productivity.

Butler Manufacturing Company

Stainless and Steel Products was sold to Butler in about 1972. We were all a little sad to see the organization that we had much affection for be folded into a much larger organization. In retrospect, the larger company would have much more financial flexibility, and provide professional growth opportunities for many of us. The top management of Stainless and Steel were not included in the realignment. I soon had a new Boss, and was given a promotion to be in charge of stainless steel tank production design, as well as development work. My life as a creative engineer was over. I would now be a full time Manager. I was moved to the Minneapolis office/shop where Aluminum tankers were built for hauling gasoline, other chemicals, and dry flowable materials such as cement and flour.

I was quickly promoted again to be Manufacturing Planning Manager for the Transportation Equipment Division of Butler Manufacturing Company. Now I had responsibilities to oversee our Material Control, Production Control, and Purchasing Departments. One of the things I did was to implement a material control system on a computer. It was an early IBM computer that utilized punch cards. I now had a hand in managing operations at our Stainless Steel over the road trailer shop, our Aluminum over the road trailer shop, and our Rail Car shop located at the Soo Line Railroad yard.

Maybe a year later, the Manufacturing Manager got transferred to the Birmingham AL Butler plant. I was promoted to Manufacturing Manager. Now I had responsibility for Manufacturing Planning, as well at all production at our three facilities, and the Industrial Engineering Department. There were now about 250 people in my group. I remember thinking, "am I good enough to lead that large group of people?". I didn't know, but I would surely give it my best effort.

With the new job came a few perks. I was given a membership in the Minneapolis Athletic club, a very nice place with hotel like rooms, fine restaurants, and a Gym. That Gym had a banked running track about 15 feet above the floor. I used it often. The Gym was also used by the Minneapolis Lakers basketball team for practice. George Mikan, Slater Martin, Whitey Skoog, Vern Mickelson, Jim Pollard, and the others played there. The unusual thing for me was the shower room. Since I am a bit over 6' 4", I often had trouble with showers that would hit me in the face or chest. At the Athletic club, the shower heads were placed about eight feet above the floor so the big guys from the Lakers could stand under them. I also had free tickets for Football and Basketball games given to me by our vendors.

Sometimes on Saturdays, I would go to one the plants to see what I might learn about what the employees didn't want me to learn about. One such day, I took Andy with me. The plant was so big that we had a bunch of Golf Carts to get admin people around it. On that day, we were riding through the plant, and Andy looked up at me and asked "dad, how did the wind get in here?" The wind was, of course, was made by the golf cart.

Another time, when I was trying to sleep in on a Saturday morning, Andy entered the bedroom. I knew he was there, and I pretended to be asleep. He lifted one of my eye lids and said "dad are you in there?"

Running

I am not sure how I got started running. Maybe it was health related. I remember running from the Dudley Ave house to the U of M golf course pretty regularly. As time went on, I wouldn't feel quite right if didn't get in my morning run. I usually ran about four miles a day at a rate of seven or eight minute miles. The first few minutes of each run was hard as my body didn't want to do it. Then 'runners euphoria' would kick in, and I would glide along easily. I ran six or seven days a week for twenty-five years. Most winter days were good enough to run outside with a little extra clothing. When it was too cold or snowy, I ran at the gym at the Saint Paul campus of the U of M or the Athletic Club.

I entered races. There were 5 K, 10 K, ½ marathons and I once tried to get in shape for a full marathon. That attempt was abandoned when I ran twenty miles one Saturday in the summer heat. I ran two loops from the Franklin Ave bridge over the Mississippi river to the Ford Parkway bridge, up one side and down the other. It was a 20 mile run. Andy followed me on his bicycle. At the end of that run I lay down on the grass, and hurt all over. I think that even my hair hurt. I thought, "why am I doing this?" There was no answer, so I went back to running lesser distance.

The races were not really races for me. The winners of the races were running five minute miles, and I chugged along at seven. If that sounds slow to you, try a few miles. Often there were hundreds, and sometimes thousands of runners in those races. I liked to start near the back of the pack, and have the sense of accomplishment in passing others.

Here is 'Lickety Split' coming to the finish line.

House # 4

This was a four bedroom split level house in Lauderdale MN, a suburb of Saint Paul. It was a step up for our family. It had a swimming pool in the backyard, and was only about 20 years old. It was in good shape, and I wouldn't have to do any fix up. We converted the swimming pool to a small skating rink in the winter time. The four kids had a lot of fun there, and often had friends come for swimming. I remember a little boy named Rusell Ebnet. He was a friend of Andy's, and would come uninvited, ringing the doorbell with a swimming suit on, and a towel in hand. He would ask, "Can Andy come out and play?" Of course, he was always welcome.

Murphreesboro

At Butler, we had two unions, the Teamsters at the St. Paul plant, and the United Electrical workers at the Minneapolis, and Soo line plants. We had contracts with both unions. Upon contract expiration the unions, especially the U E would threaten strike to further their bargaining position. They seemed to think that they couldn't get a maximum contract offer unless they struck. The Teamsters, on the other hand bargained hard, but strikes were few, and very short. The Teamsters had an attitude that they would bargain hard at contract time, but then work with management during the contract. They knew that by working with management, the customer would be best served, more money would be made, and that money would be available for the next contract negotiation. The result was that we endured a few strikes at the Minneapolis plant. Strikes are very disruptive to business. When we had a strike, production was halted. Our customers that had tankers in process did not get delivery as they had been promised. The strikes would set production back three to ten weeks.

So what to do? We needed an alternative plant that could be kept open even during U E strikes. We did a lot of research looking for a good place to build a new plant. The new plant would have to be centrally located for easy transportation in and out of it. It would have to be in an area that had a ready supply of employees, and it would have to be in an area where 'Union Free' shops were welcome. Murphreesboro Tennessee was the answer. We talked to the city fathers (Mayor and City Counsil) The City wanted new Manufacturing plants. They even asked us about our attitude towards unions. We found that they thought a lot like us. The city also agreed to help us financially. We would buy a plot of land in what was to be a future 'Business Park', and they would pay to have raads, sewer, water and electricity brought to the site.

Now the fun began for our manufacturing people. We would design a plant that was capable of building over the road tankers and railcars. We could design it from scratch, and not be burdened with having to fit our production ideas into an old existing building.

Before the building was finished in 1975, we worked with the Tennessee Vocational Tech people to train a production staff. We identified six people (a superintendent, and industrial engineer, and four foremen) to be transferred to Murphreesboro as the core of our new staff. The state paid to rent a facility for training employees. We started by interviewing prospective employees, and described to them what our plan was. They would go to classes put on by our core six people. The classes were to be held in the evening, four days a week for three hours each for four weeks.. We taught them blue print reading, sheet metal forming, welding, and assembly. During the class time the prospective employee would get to know us, and what would be expected of them as our employees. And, our core team could evaluate each prospective employee. We had twenty people in each class, and repeated the classes four times to get a beginning production staff.

That plant was up and running quickly, and after about a year it was profitable. We treated those new employees very well. They were paid above the average wage for similar work in the Murphreesboro area. They had all of the same benefits as our office staff. We treated them with respect.

I did a lot more traveling in those days. I needed to make a trip to Tennessee every few weeks.

A sad memory was of the death of one of our new employees. He was the head Maintenance man, and a good man. I had talked with him several times.

The plant was designed to be able to take full rail cars in and out of big doors. Those doors were 12 feet wide, and 16 feet high. They were driven up and down by electric motors. The control box for one of those doors was erratic. That door would sometimes open when a train would pass on the nearby main line. Its control was too sensitive. Our head maintenance man was working on it one day. He worked from a hydraulically operated lift that could take him to the top of the door. The lift had stabilizer feet that extended in four directions. He brought the lift near the door, and energized the hydraulics to lift him to the troublesome motor. He had inadvertently put two of the stabilizing feet near the frame of the door. As he tried to adjust the mechanism, the door was triggered to go up. As it went up, it caught the supporting feet and tipped the whole lift with him at the top. The lift fell to the ground, and our man held tight to the upper guard rail as it went down. His head cracked against the concrete floor.

I learned of this within the hour, and arranged to fly there the next day for the funeral. It was a very sad time for all. He was in the process of getting a divorce. His wife filed against him. At the funeral, which was an open casket ceremony, his wife stood near the casket for the entire ceremony. She would not let the funeral director close the casket. There was much crying.

The person that I most respected at the new plant was the superintendent that we transferred from the Minneapolis. His name was Jim McCormack. In conversations with him, we reviewed the performance of the Minneapolis plant. Aside from the problems with the union, there was an attitude on the shop floor by many employees, that they didn't need to work hard, and that the big Company would always be there for them. Jim told me that I had to convince them that lackadaisical work ethic was not acceptable, and that in order for the plant to be viable long term, we had to be competitive with the other companies that made tankers.

Productivity Turnaround

I developed a plan to get all of the employees, shop and office, on board. We would take time away from the shop routine to educate all of the people in how the business works, and what we needed to do to be competitive. We would take them in groups of forty people, and set up a classroom environment.

The first session began with a warm greeting, and telling them that we wanted them to see all aspects of the Company. I brought out ledgers, and printouts of cost structures of our products, overhead costs associated with production, costs of supporting departments, building costs. I told them that any information in those ledgers was available for them to peruse. The only thing that I wouldn't discuss was salaries of the office staff.

I expected that the Union would give a lot of flack about how management made it impossible for the workers to be productive, and if they were paid more, then they would be more productive. No Union flack was put forward.

I made a big drawing of a typical over the road trailer tank. That drawing was six feet wide, and four feet high. I used it as a chart to breakdown all of the costs, and profits, and taxes that went into a typical tank. Each component occupied a section of the tanker, and was colored differently to make it stand out. The width of the section was proportional to the percentage of total selling price that the section represented. Sections represented were; Materials, direct labor, indirect labor, engineering overhead, sales overhead, accounting and clerical overhead, factory supervision overhead, building costs, warrantee costs, local state, and Federal taxes, and finally net profit. People had heard that our profit margin was about 20%. It was actually more like 4% after tax. In bad years, it fell to become a loss.

We discussed the components of each section, and welcomed questions. The point of all this was to show that in the competitive environment of the tanker business, we lost some business due to our inability to reduce selling price below that of our competitors. If we could take 5 or 6 percent out of the cost structure, our sales people would be able to reduce selling price, and get more business. More business meant more job security for all of us. More business also meant that the fixed overhead portion of each tanker would be spread over more tankers, and the resulting spread would further allow again, lower total cost per tanker, and allow a further reduction in selling price. More business meant that we would have more flexibility for future pay increases.

As we talked about each section, it became apparent that some of the cost sections couldn't be reduced: Materials were already being bought at minimum prices. Engineering was a necessity. Taxes were unavoidable. Perhaps some small savings could be expected in other sections, but the biggest potential was in direct labor. Now everyone in the room knew that there was plenty of goof off time on the shop floor.

I put it to them that we didn't expect people to work harder, but if they worked smarter, and steadily throughout the day, we could achieve cost reductions that would result in our ability to sell more tankers.

Also, there was the cost of inventory. We talked about the company borrowing operating capital from the banks. If we had $1,000,000 in any kind of inventory, and we had a 5% rate from the bank, then we had an additional $50,000 per year to be absorbed in the cost of our tankers. We always tried to minimize the amount of dollars we had in inventory. That inventory could be nuts and bolts, or it could be in stainless steel sheet waiting to be turned into vessels, or it could be finished tankers ready for sale.

Those meetings took place at the rate of six per day. I was able to meet with all employees (including office people) in one day. The meetings took two days in the first week. I was exhausted at the end of each round of meetings, but I felt good about the communication that we did. We would later meet in the same style whenever good communication of a situation demanded it. The employees would be a part of the team.

We talked about what we did to avoid layoff of workers during a time of poor business. We had a number of 'Standard Tankers' that we build on speculation that customers would buy them. We knew that those tankers would always have a market. My philosophy was that in times of good business, we would let the inventory of 'Standard Tankers' go down, and that in a period of poor business, we would let the inventory of 'Standard Tankers' go up. Thus we would use our workforce to make mostly "Customer" tankers in good times. In bad times we built whatever customer tanks we had orders for, and we built for inventory. In this way we could keep a pretty steady work force without having to lay people off during slack business times. Nonetheless, sometimes it became necessary to lay off good employees. I very much hated to do that. When it had to happen, I personally would meet with the effected people, and tell them of the lay off. I would explain our philosophy, and apologize for not having it work. I felt better, but they were out of a job.

We wanted to make the transition to improved productivity as easy as possible. We set up an interview for each employee on the job with an Industrial Engineer. Each employee was asked what we could do to make his job easier, and what tools he would require to accomplish that. The employees were very appreciative to have their opinions asked. From the lists of Jigs, Fixtures, and better hand tools that came out of this, we bought about 95% of the ideas.

I had another chart for the meetings that showed the connection between Management, Employees, Customers, and the Owners. It consisted of three large circles, each of which was named; Customer, Employee, and Owners. In the center of those circles was a circle entitled Management. I told each group that management's responsibility was to make each of the three groups happy. Further, if we were able to make Customers and Employees

89

happy, that then the happiness of the owners would automatically follow.

At the same time we wanted to brighten up the work place. Each of these plants was 60 or 70 years old, and pretty much had never been painted on the inside. We arranged to have the entire ceiling and walls of each department cleaned, and painted white. The result was that the buildings were more cheery, and the increased light from the reflective white surface made the work areas brighter.

The attitude in each shop started to get much better.

We had a way to measure worker performance. We called it 'efficiency rating'. Each task on the shop floor had a standard number of hours that we expected it to be finished in. Those standards were arrived at from historical data. Thus if the standard was 10 hours to make a batch of side channels, and the crew took 9 hours to build them, they got an efficiency rating of 90%, a good thing. Conversely if they took 11 hours, they got an efficiency rating of 110%, a bad thing. At the end of each day, and from time clock records, all of the efficiency ratings were tallied together. On the average day for many years preceding the 'Productivity Turnaround' the efficiency rating was 100%.

I wanted to communicate directly with the shop employees on how they were doing. I made a big chart to be placed in an area close to the time clocks. On that chart, I recorded the efficiency percentage for the day. The people all watched that number. They felt part of the team and wanted the efficiency to show that the pace of work had improved. I watched the daily numbers, and was gratified to see increases in productivity.

Credit for most of this should be given to our Foremen, and Industrial Engineers. Without their support the plan would not have worked.

We soon reached an efficiency improvement of about 20%. Wow, that was a very big deal. And, the shop people were not working harder. They worked a bit smarter, and a lot steadier.

Soon thereafter, the sales manager, Bill Patten, came to me and asked if the new savings could be passed on to the customers in the way of lower prices. I was confident that the savings would last, and told him that if needed in competitive situations, the prices could be reduced by up to five percent, and we would still meet out profit goals. Well, we got more business, the overhead was spread over more tankers, and all prospered.

We had our first 'bomb scare' one day about mid morning. Someone called into the switchboard, and announced that there was a bomb in the Plant that was set to go off within the hour. We quickly decided to evacuate the building. Getting a hundred people out in a calm manner was not easy. After evacuation the police bomb squad gave the plant a thorough inspection with sniffer dogs. There was no bomb, but production was disrupted for about two hours. This repeated itself two more times, before we decided to call the next one a hoax. Bomb scares kind of caught on in the area. Other buildings went through similar shutdowns.

Boat #6

It was a 1961 28-foot Rhodes Ranger. It was build in Heenvleit Holland. It was fiberglass, and had a 10 HP Volvo Diesel one-cylinder engine. The generator was also a starter. If you moved a switch to the right, it was a starter, if you moved it to the left, it was a generator. She slept four in modest bunks, and had a small galley with an alcohol stove. It was a very good sailboat, and was worthy of going in the open ocean under any conditions. We spotted it parked along side of Highway 12 leading into Wayzata MN. It was in an estate, and was poorly maintained on topsides. We bought it at a very reasonable price, and moored it on Smiths Bay of Lake Minnetonka at a buoy owned by the Smiths Bay Boat Works. Wow, this was fun! I had her documented and registered with the U S Coastguard. Her new name was 'Cathleen', named after my wife.

We fixed up the worn Mahogany topsides wood, and had many pleasurable hours of cruising on Lake Minnetonka. Son Andy was right there with me to help with the fix up.

Two years later, the boat works decided that the only boats on moorings in front of their place would be boats that had been purchased from them. This was in the winter time, and we had to arrange to have our boat removed from their property very soon. I had visited the Apostle Islands earlier, and was determined to moor the boat there. We arranged for transportation up the highway to Bayfield Wisconsin. Bayfield was only three miles from the Apostle Islands. Our new berth for the Cathleen was at the Madeline Island Marina. We had a slip with power and water at dockside.

Lake Superior is big and cold year round. Its depths go to about 1300 feet. It holds 10% of all of the fresh water in the world. The wind and waves can build up in a hurry, and provide challenge for any mariner who thought he was up to the task. Lake Superior temperatures are in the forties in the summer.

We explored Western Lake Superior from Duluth to the Keewanaw Peninsula. Our big trip was one that I knew might get me ready for ocean adventure. We were to go from the marina at Madeline Island to Isle Royal Michigan. That would be a trip of about 120 miles, and would require an overnighter on the big lake. We shortened the trip a bit by starting from an anchorage on Outer Island of the Apostles, and heading for a Marina in Grand Marais MN. This cut it down to about 95 miles. To prepare for the trip I needed some way to determine our location at any time. We would be out of sight of land for an extended time, and would not be able to use dead reckoning. We had no GPS, Loran C, nor RDF. I read about using a conventional hand held radio as a locator. It was really very simple if you had the guts to trust it.

Our little radio had an antenna inside, and could be turned to the best angle to get good reception from a shore AM radio station. It could also be turned to get a 'null', or, no signal 90 degrees from best reception. When a null was detected, you knew that the little antenna was lined up with the shore station. Now, I recorded the Magnetic compass direction to the station. I knew where the station was located, by listening to the broadcast. Next I drew a line on the chart from the station location and out on the lake at the recorded angle. Our boat had to lie somewhere on that line. Next was to get the direction and location of two or three more stations. A line was drawn on the chart from each station location and out into the lake. Viola! All of those lines intersected at one point, and that was our position on the lake. We left early one morning from the anchorage on Outer Island. We sailed through the day, and all night. At five knots, we should reach Grand Marais in 95/5 or about 19 hours. It was a long cold night, but we were happy to see the Minnesota shore the next morning.

Exploring Isle Royal National Park was surreal. There is no place to stay on the island, and it has no electricity, and no roads. Yet it is pretty large, maybe 20 miles long with lots of little inlets suitable for anchoring. We were in the wild, and saw lots of wild life. As part of our preparation for the trip, I took a small Styrofoam cooler, filled it with selected cuts of meat, and topped it off with water. I then put the whole thing in the freezer and froze it solid. The idea was that the insulation would hold for a time, and then the ice would gradually melt keeping the meat fresh. Wrong! Unbeknownst to me, the Styrofoam developed a crack in it from the freezing process. Then, as the little cooler warmed up in the lazarette of the boat, the water inside melted easily and dripped out of the crack. When it was time to cook up some nice little steaks, we discovered the cooler was warm, and the contents were rotten! Oh well, we had back up food in tin cans.

One morning, while we had breakfast ashore, I brushed my teeth in the lake. About that time a Park Ranger came by in his high powered motor boat. His base was 20 miles away in Grand Marais, and he made the trip regularly to 'protect' the island. He saw me brushing, and gave me a lecture about "polluting" the lake. I, of course said, "Yes sir". As he roared away from the anchorage in his boat, the exhaust from his two cycle engine spewed foul products of combustion into the lake. Now, that was pollution!

Cathleen

Divorce

One day at the office, a Process Server visited me. He served me with divorce papers.

It wasn't a total surprise since there had been troubles in my marriage to Cathy for a few months. Nonetheless, the abject finality of it was very disturbing.

I won't try to assess blame for the situation. The adults involved possibly had not chosen their mates correctly, and/or the adults involved didn't do what was necessary to keep their marriage healthy. In any event, we were to go down the difficult road to a final divorce.

The huge injustice in the divorce was the effect it had on four beautiful and innocent children. They were not able to understand what was happening, and what caused it. All they know was they were losing their family. They were made to be witness to bickering, and blaming. I am sorry for the part that I played in the irrevocable situation.

I was ordered by the court to leave our home. I took up residence in an apartment in New Brighton. The children came over often, and liked to swim in the big pool there. We would cook in the apartment or go out for meals. We were together for too short a time.

I had an Attorney named Larry Katz. He was a small (5' 3") Jewish guy that was very bright and articulate. We were in a custody battle. I wanted custody of those four beautiful children.

I believed that I was the best person to carry on with parenting. There was much investigation by the family court, and some visits to Psychologists, and Marriage counselors. The outcome was that the court found me to be the most competent parent. I was told to move back into the house on Carl Street, and Cathy was to vacate the premises. All of this was very hard on the children.

Back in the house, I had to arrange for care of the children when I wasn't there. I soon hired a Nanny who was a student at the University. She would come on weekdays before the children got out of school, care for them, do some light housekeeping, and cook the evening meal. The agent of the family court, a Mrs. Duddleston, oversaw the whole scenario.

Things were getting to be back to normal.

I had to sell the Cathleen boat in order to make ends meet. She got a new home on the Saint Croix River near Hudson WI.

A New Family

A year or two later, I started dating LaVonne, who had also been divorced. Some of our kids had gone to school together in Lauderdale. The love bug bit, and we planned to get married. All of the Children, Clark and Carol Misner, and Colleen, Mary Claire, Andy, and Sara Olson were good with this. We decided that we needed to start out our new family in a new house, and not move one set of children under the roof of the other set. All of those children were to be 'our children', not mine and yours. The year was 1977.

We found a very nice house on Pike Lake in New Brighton. We made a deal to buy the house, and then brought the children into the deal with a small bit of deception. We wanted all of those kids to want to live in that house, and be part of the new family. One day we took all of them to inspect the house as if we were still looking. They were all ecstatic about the prospect of moving to that house on the lake. They okayed the purchase of the house. We were starting to make a new family bond.

The Pike lake house had five bedrooms one of which was unfinished. The lower level had the one unfinished bedroom, and an unfinished bath and ante room. It had no deck when we bought it. The work of building a deck, and finishing off the interior was something that I easily did, and enjoyed. We now had six children in grade school and junior high. I remember Andy and Carol being on the same baseball team. It was a boys team, however they found room for Carol too. As I recall, Carol turned out to be the best player on the team.

1320 North Pike Lake Court, New Brighton MN,   circa 1985

This was a marvelous house for a big family. It had 165 feet of frontage on Pike Lake. The living room was large, and had a picture window overlooking the lake. The family room adjoined the kitchen and had a sliding glass door that over looked the lake and served as access to the big deck that I built after we moved in. I mounted a Webber charcoal grill on the outside of the deck railing. It made cooking easy. Summer time deck meals were very nice. We even used that grill in the winter time for roasting turkeys. We would shovel the snow off of the deck and do the barbecue thing.

The lower level of the home had two bedrooms, and bath, a laundry/shop room, and a big recreation room. It had a sliding glass door that opened onto a rolling yard that led to the lake. We bought a full size slate pool table for the rec room. The pool table also served as a ping pong table when the five foot by nine foot green top was clamped onto it. One thanksgiving we invited our two extended families for dinner. We extended the ping pong top by adding another piece of plywood held up by a saw horse. I think we had 16 people seated at the table. The logistics of serving the food was a bit complicated.

There was a small dinette area adjacent to the kitchen. We needed to have room for all eight of us to have meals there. The space was just big enough. I built a table from a solid core wooden door 36" x 80". I gave it a frame made of cherry wood. I thought about how to build chairs that were strong, and small enough to fit around the table. I always liked Danish Modern furniture design. I wanted to make the chairs without using screws or nails. Dowel pins were the answer. I made a prototype chair from cheap and soft redwood. I experimented with different shapes, and when I was happy with the design, I built eight chairs out of maple. Those chairs are still with us today, 38 years later. The table is now a fold-up-from-the-wall work bench in the garage.

I built cubicles for the laundry room. There were eight of them about 12 inches deep, 12 inches wide and 16 inches high. We were of course always doing laundry with our large family. There was a laundry chute in the room that delivered dirty clothes from upstairs. When the clothes came out of the dryer, they were sorted by owner, and put into the corresponding cubicle.

We had a small black dog named Inky. Actually it was Clark's dog, and Inky slept with Clark. That dog had an every day game with the squirrels. Inky would lie in the grass 20 or 30 feet from the squirrels tree. The squirrel would come down out of that tree, and slowly approach Inky. When the squirrel was about five feet away, Inky jumped up and chased the squirrel. The squirrel always made it back to the tree first. It would go up the tree part way and then let out 'squirrel talk' to tell Inky who won the race. That little dance was repeated over and over again.

That little dog did us a very big favor. As she got older, and her health was failing, she began 'leaking'. We knew the end was near and dreaded what we would have to do. Inky went out in the street and was killed by a car. She saved us that final trip to the Vet. We were all very sad.

When we first moved in to the Pike Lake house, we planted some birch trees and two weeping willow trees. The willows were planted on the western border of the property to provide some separation with the neighbor. They grew to be large trees in a few years and were beautiful. One day we got a letter from a lawyer demanding that we remove those two willow trees. It seems that the neighbor had an easement on the end of our property for the purpose of access to the lake. The trees supposedly blocked full access. They never used that easement. Why didn't the neighbor come to us, and discuss the situation? What to do? We had a think on the problem, and decided to do nothing. More letters came over the next few months, and we took pleasure in knowing that every 'lawyer letter' cost our neighbor about $100. Eventually, I responded, and made a deal that we would leave those trees there until we sold the house.

The house also had a nice lilac bush outside of the master bedroom, and the bedroom below it. In the spring time, we would leave the windows open, and the marvelous lilac fragrance would waft into those rooms.

The Pike Lake house came with an old motor boat with a 25 horse engine. It was a pretty good ski boat for the kids. Also, in the spring, after we moved in, I found an Aluminum canoe that was abandoned and had been allowed to freeze in the ice that winter. The Canoe was badly bent, but after some careful straightening, it was suitable for us.

The kids used all of the boats on the lake. The lake was pretty round, had no inlets or bays, and was about three blocks in diameter. From our living room or deck, we could see the whole lake. We were able to watch the kids as they played on the lake with boats, or on the swim raft.

The family room had one wall that was faced in brick. It also had a nice fireplace in the center of it. Every year in the late fall we journeyed to Hand Lake with a big trailer to get firewood. Brother Mike and his in-laws owned a big section of lakeshore there. A lot of trees in the adjoining woods would fall and just clutter up the floor of the woods. We brought chain saws, and cut up those birch and oak trees. We split the wood on the spot with axes and wedges and loaded it onto the trailer. We stayed there over a night or two, and had much fun just being there. At home all of that wood was neatly stacked inside and out side of the garage, and near the family room. All winter long we would build a nice fire almost every night. The splendid ambiance of a crackling wood fire in the middle of winter was always cherished.

Wintertime brought ice to Pike Lake. Starting in December, the ice thickened up. When it got about two inches thick, it started to crack. All good lake people know that when the ice cracks it is safe to walk out on. We skated all over that freshly frozen and smooth ice. That was fun. Then the snow would start, and the kids took to shoveling off a rink so they could continue skating and playing hockey. Each time it snowed they shoveled. With bigger snowfalls, the rink got smaller as the shoveling load increased.

We also had two old snowmobiles. They were fun to ride on. We could cross Silver Lake Road and ride over that lake and its adjoining terrain. Those snowmobiles were often in need of fixing, something that wasn't much fun outside in the cold.

During the second winter with the new family, we rented a motorhome and drove it to Florida for Christmas vacation. It was big enough for all of us to sleep in. I remember driving it myself straight through to northern Florida. In northern Florida I parked it and got 2 or 3 hours of sleep. Our first stop was in Fort Lauderdale where LaVonne's dad, wife, and half brother and sister lived. We toured Disney world, Sea World, and Bush Gardens. Gee we had fun.

Reunion

My mother had five sisters, and each of those had families. They were scattered all over the United States at that time. We talked about having a reunion, and planned it a couple of months ahead of time. Colleen was quite good with calligraphy and prepared the invitations. The whole clan responded, and came. We really had a good time reminiscing, playing on the lake, and enjoying good food and drink.

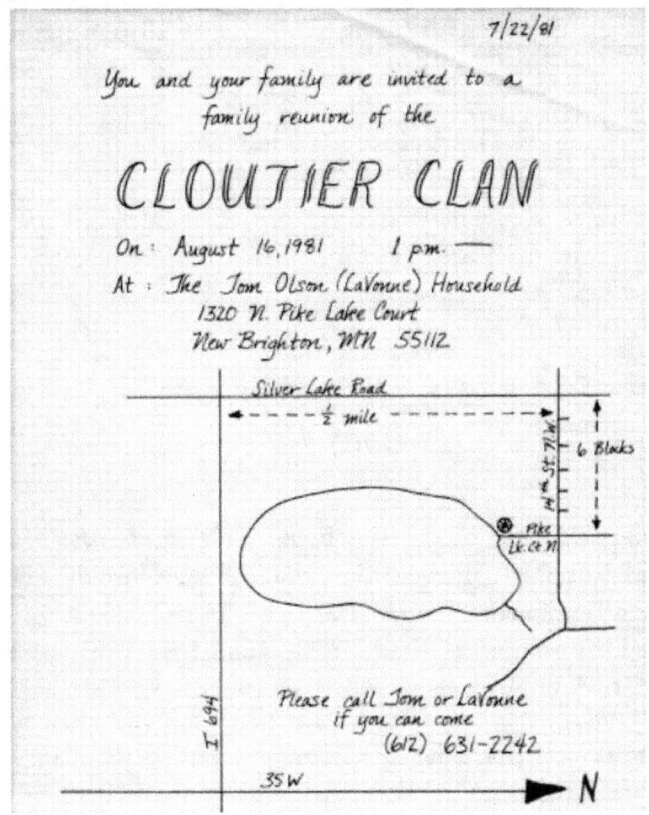

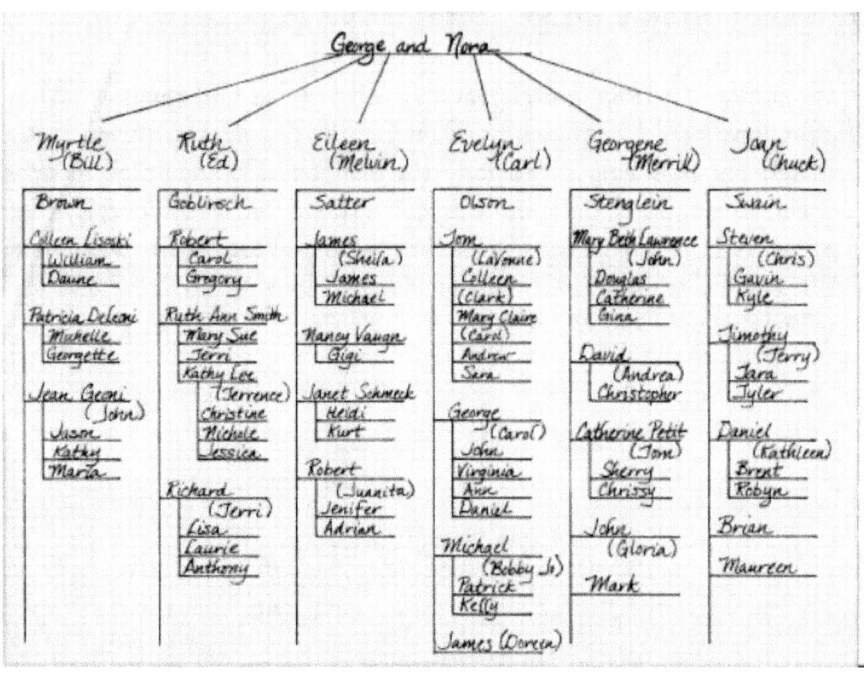

Boat #7

We (I) needed a sailboat. We bought #7, a 16-foot Hobie Cat That boat was fast, it could do about 15 mph in a 10 knot beam wind. We all had much fun on that boat. It was also a bit tippy. I remember the very first time I took it out that I pitch poled, and capsized it. A catamaran is sensitive to weight distribution fore and aft. When it is on a reach or a beat and the crew weight is too far forward, the leeward bow has a tendency to bury in the

106

surf. When that happens, the whole boat is pitched forward, and over. What an awakening. I thought that all of my sailing experience would let me 'captain' that Hobie Cat. We learned to keep the weight balance a bit abaft of beam.

That boat was so fast that I decided to try it at pulling water skis. At the time, Andy weighed less than 100 pounds, and was a pretty good water skier. We put a tow rope behind the Hobie Cat and set out across the lake. Andy was able to get up on the skis. He would go along pretty good until the wind died. He then sunk into the water, and would pop up later when the wind picked up. He was a good sport about it.

Later, when we owned the Vonnie T and were in the final stages of planning our big trip, I put an ad in the paper to sell that Hobie Cat. I was up at Lake Superior that weekend doing some work on Vonnie T. When the people came to inspect the Hobie Cat, LaVonne told them that she did not like it as it was too fast. Well that turned out to be just what the prospective buyer wanted to hear. The boat sold quickly.

16 foot Hobie Cat

Butler / Transportation Equipment / Donaldson

For reasons unbeknownst to me, Butler decided to sell the Transportation Equipment Division. Some things were said about Butler wanting to concentrate on its core businesses of Steel Buildings and Grain storage.

I was a key guy in the sale. I knew the most about the facilities we had, and would show prospective buyers through them. I kept the coming sale a secret until it was a done deal.

The sale was made to a surrogate of A M General Company. They bought the business with the exception of the Murphreesboro plant, and the rail car business. A M General was pretty big in vehicle production. They built military Jeeps, trucks, and some civilian cars. The new owner was a man names Archie Albright. He had other business interests and wasn't able to be on the scene all of the time. An A M General manager was brought in as the General Manager of the new business.

I had always respected every boss that I ever had until now. The new guy was a blustery graduate, and football player from the Naval Academy. If he wasn't sure or didn't know something, he would often bluff his way along.

Professional 'Recruiters' also known as head hunters ply their trade in Corporate America by luring professional people out of their present positions, and into the position their client has open. They charge exorbitant fees for their services, perhaps as much as a years equivalent of the recruit's salary. Professional people often receive calls from the 'Recruiters' perhaps monthly, or at least a few times a year. They describe a new employment opportunity, and try to get you interested. I pretty much despised these guys, and brushed them off quickly.

After trying to work with the new company, now called Transportation Equipment Corp, for about a year, I got a call from a Recruiter, and gave a yes to the proposal. This new company was Donaldson Company headquartered in Bloomington MN. One small concern was that my brother Mike worked for Donaldson. Mike worked in a different division, but I didn't want to do something that might in any way hinder his career. I discussed the offer with Mike before accepting it. He was excited about having me be a part of Donaldson.

I became the plant manager for the Torit division of Donaldson. I found the new plant to be running pretty smoothly, and relegated my self to keeping it that way. No major change would be necessary.

Some time later, I was challenged by Donaldson to take over management of its Eagan MN facility where air cleaners for heavy equipment were made. There were labor problems in that plant that were fermented by the union, the United Auto Workers. Management, over a period of time, had kind of 'thrown the reins out of the buggy'. Things were done to placate the Union, instead of furthering the prosperity of the plant.

On day one at the Eagan plant, I was rearranging some nice office furniture that I had brought from the Torit plant. The union filed a grievance on me for usurping the work that should have been done by the maintenance department. It was their way of telling me who was running the plant. It was a grievance born from 'chicken droppings'. This was going to be fun, I thought 'Game On'. I ignored the grievance, and went about learning about my new responsibilities. Shortly I could see that the Union was way too influential. Productivity suffered as a result of it.

I was determined to win over the work force just as I had done at Butler. We arranged meetings for all the plant people, and I explained how the business worked, and that the real opponent was our competitor. Just as at Butler, we got no flack from the Union. We, again, gave them immediate feedback by the way of productivity charts. We were able to improve attitudes, and did achieve a significant productivity increase.

We ignored the Union.

One of the nicest things happened next. The president of Donaldson, Bill Hodder, had heard of what we did, and wanted to see it first hand. We arranged an afternoon for his visit. We toured the plant, and I introduced him to some of the plant people. We then adjourned to the conference room where I had set up all of the charts, and easel drawings that I used in the presentations to the plant people some months earlier. I went through the same presentation with him. Bill and I hit it off.

Donaldson had a division that made a business of providing filtration equipment for the U.S, Canadian and Israeli military. That division needed some new direction. I was promoted to the position of Director of Defense Products. Now, I had overall responsibilities for Sales, Engineering, and supply of the products. Those products included air filtration equipment for army trucks, tanks, and aircraft. We also sold filtration equipment to be used in buildings or tents that were sealed off to prevent the inhabitants from being exposed to particles distributed by Chemical, Nuclear, and Biological warfare.

That filtration equipment was to be used where an attack had happened, or where one might occur. The equipment used barrier filters to remove the foreign particles from the air before it entered the building or enclosure. The air was also passed through specially designed charcoal filters that could remove the non-particulate bad stuff. We dealt with government contracts, and a huge military bureaucracy.

Here is another small piece of history that I was involved in. We made 'Particle Separators' for gas turbine engines that were mounted on military helicopters. Now, recall the 'Iranian Hostage Crisis' that occurred during the presidency of Jimmy Carter. It started on 1/20/1981 and lasted for 444 days. It was during the Iranian Revolution. The U S had an Embassy in the capital city of Iran, Tehran. That Embassy was taken over by a group called the 'Muslim Student Followers of the Imam's line'. They held 52 US citizens hostage. The hostage taking consumed more than a year of Jimmy Carter's presidency.

The U S planned an attempt to rescue the hostages by bringing in helicopters to carry them away. I believe this was a huge planning effort.

A ship carrying the helicopters, some fixed wing aircraft and other military hardware went to a location on a desert beach. A small base was erected to facilitate launching of the helicopters. In order to carry maximum payload, the helicopters were stripped of armament, and anything that was deemed heavy and unnecessary. One of the things they removed were the Donaldson

Particle Separators that protected the helicopters' gas turbine engines.

On the day the mission was to begin, the helicopters took off. They kicked up a big storm of desert sand. That sand was ingested by the helicopter engines. The engines failed because they were not protected by our particle separators. The mission was aborted.

Now, had the particle separators not been removed, the mission might have worked. Then Jimmy Carter would have been the hero for being the Commander in Chief that oversaw the operation. Jimmy Carter would have likely been elected to a second term.

As it turned out, the Iranians released the hostages shortly after Ronald Reagan became president.

A few years ago, I met an ex army officer at a golf course. We talked about random stuff like golfers do. It came out that he was on the ship that delivered the helicopters to that remote desert location. What a small world!

Real Estate

The Bedford Street duplex was our first investment in real estate. I had read a book 'How I Turned $1,000 into a Million in Real Estate' by William Nickerson. It was a simple formula. Buy a building in a good neighbor hood that had low rents because of poor maintenance, and fix it up. The fix up warranted higher rents, and the higher rents increased the capital value of the building.

We bought and sold several properties over several years. When we lived in the Pike Lake house, a friend of ours who happened to be a real estate broker brought a proposition to me. He had found a 218 pad Mobile Home Park near Rochester MN named Oronoco Estates that was for sale. The numbers indicated that it would be a good investment. Three of us planned to buy it. I well remember not having enough money for my share of the down payment. In those days we did banking at the 1st State Bank of New Brighton, which was near to the Pike Lake house. Banks like that no longer exist, having been gobbled up by larger banks. The First State Bank of New Brighton was independent, and operated with its own President and Board of Directors. I had met the president a few times, and thought I would ask him for a loan. I needed $50,000 in addition to our savings. LaVonne and I had pretty good incomes, and we had a nice balance sheet that included real estate equity. The president looked over my situation, and agreed to recommend the loan to his Loan Committee. That loan was unsecured. The Loan Committee accepted the deal, and we were off and running.

We had a lot of our money invested in the partnership that owned Oronoco Estates. The partnership deal that was struck called for the Broker/Partner to get a disproportionately higher share of the ownership than would be ordinarily be associated with his down payment. In return for that he agreed to transfer to me all of his depreciation for a period of five years. This was a perfectly legal transaction, and it allowed LaVonne and me to use the extra depreciation to offset some of the income from our jobs. The net result was that we paid no income taxes for a few years. We got some cash flow from that investment and held it for 30 years. When it was sold, we had fully depreciated it, and would owe the government for the gain.

We wrote the IRS a very big check for income tax in the year of the sale.

About two years after buying the mobile home park, we were ready for another venture. We bought an Office Warehouse in Fridley MN. I directly managed it myself, and dealt with Tenants, and building upkeep. That building provided us with some more cash flow. We held that building for 28 years before selling it.

These investments gave us enough financial independence for us to see our way to an early retirement.

Boat #8

The Vonnie T

LaVonne and I chartered boats on various occasions in the Virgin Islands in the Atlantic Ocean. We enjoyed those vacations a lot. Some of the kids accompanied us on those trips. When it was winter in Minnesota, we went to the Caribbean. We got some more experience sailing on big waters, and seeing magnificent places.

Early on, when I was still in my twenties, I had two lifetime goals. One was to run in the Boston Marathon, and the other was to sail around the world.

As was earlier mentioned, I gave up on the marathon goal. The sail around the world goal comes to many sailors, and I thought 'why not me?' LaVonne had always been an eager traveler, and was open to adventure. We formulated a plan to start our sailing adventure when our youngest, Sara, was off to college. By that time the other children were mostly doing their own thing away from home. Our time frame would be to buy a boat in 1985, and begin on our odyssey in 1987.

We started searching for our dream boat. We looked on the east coast, particularly in the Chesapeake, Florida, and Hawaii.

We were having a hard time deciding what type of boat would work for us. One day when we were bouncing from Marina to Marina in the Chesapeake, we got the idea of hiring a Marine Surveyor. A Marine Surveyor is a person with higher education in boat construction and boat handling. They have a lot of experience around boats. They are normally hired to survey a boat at the time of sale. They come aboard and inspect every nook and cranny of the boat being sold. Then they write an exhaustive report on the condition of the boat, and also offer an opinion of the boats value.

We picked out a Marine Surveyor, and paid him to talk to us about what boat might be best for us. It was a very worthwhile few hours. We were now looking for a boat that came from a yard that had a good reputation for seaworthy construction. We were looking for a boat that was large and comfortable. We were looking for a boat that two people could handle at sea. We were looking for a boat that a person 6' 4" could stand up in.

Here is a little lesson on who not to trust. We found a boat in Fort Lauderdale, that we liked very much. We made an offer, which was about 20% less than the asking price. We deposited earnest money with our offer, and went home to Minnesota thinking we would come to some selling price between our offer, and the asking price. I thought we had that boat. The next day, the broker called to tell me that they had sold the boat to someone else. What? If someone else had come along, and offered more than we had, why were we not contacted to make a counter offer?

I am very sure that the broker presented our offer to the owner, and the owner accepted it straight away. Now, the broker knew the boat was worth a lot more that our offer. My best guess is that the broker bought it himself at the bargain price, and turned a nice additional profit for himself. The same thing can happen in real estate transactions. The brokers know the values, and some lack integrity.

We were disappointed, but determined to find our boat. In October of 1985, we found a 50 foot 1981 Gulfstar Ketch in Fort Lauderdale. Its name was 'Cottons Harvest'. A commodity broker who went broke had owned it. The bank had repossessed it, and it sat dormant in a canal in Fort Lauderdale. The topsides varnished coamings, hatchway, and gunnels were in bad shape. The beautiful teak was still intact, but the varnish was peeling off. Below decks, she was pretty much immaculate. She had reverse cycle air conditioning/cabin heat. She had a full cockpit enclosure made from Isinglass and Sunbrella that was later to keep us warm and dry in some pretty nasty weather. The engine had only 250 hours on it. We knew immediately that this was our boat. We made the deal, and took out a loan to pay for it.

We redocumented it, and gave her the name 'Vonnie T'. LaVonne's father had called her Vonnie when she was young, T stood for Tom, and the name sounded like a tug boat. All good!

Vonnie T at anchor in Cook's Harbor
Morea,   French   Polynesia

We returned to Fort Lauderdale that Thanksgiving. I remember having Lavonne's dad, George and his family aboard for our first venture at sea. George helped us look for boats in the Fort Lauderdale area. Boy, he was proud as a peacock sitting in the cockpit as we motored out through the seaway and into the ocean.

At Christmas time we took the boat to The Bahamas for the first time and thence to Georgetown SC for the winter.

Now, to get our new girl home....

The voyage from Georgetown would take us up the eastern seaboard to New York Harbor, the Hudson River, the Erie Canal, and the great lakes. I figured we should make that trip of about 2200 miles in about three weeks. We would need to sail pretty much around the clock, and would need crew. We arranged for six friends in crews of two to meet us at various places along the way.

The first leg was out to sea and in the Gulf Stream. The Gulf Stream is a huge current that flows along the eastern seaboard from south to north. It actually starts in the Caribbean, and ends near England. The Gulf Stream is five to twenty miles wide, and very deep. It moves at 1 to 3 knots (a knot is a nautical mile, about 1 1/9 land miles) per hour. Now here is a fact for you. The Gulf Stream contains 25 times more water that all of the rivers in the world.

We made that first jump from Georgetown to New York City in the Gulf Stream, and caught many 10 to 12 pound blue fish along the way. Now that is good eating. We found a Marina in New Jersey, right across the Hudson from Manhattan Island. We were able to change crews there. I know that Northern New Jersey has a reputation as being a pretty tough place but this marina welcomed us. I asked if they could accommodate us for the night. The owner asked where our boat was. I told him it was at his gas dock. The gas dock was closed for the night. This marina was mostly for small boats, up to about 30 feet. We were 50 feet, and drew 6 feet of water. The owner said we could stay there for the night if we promised to be gone by 7 the next morning. He would not accept any payment for the over night berthing. That was New Jersey hospitality! As night fell, we fixed our evening meal of Bluefish, and sat in the cockpit viewing Manhattan and watched the full moon come up over Manhattan. Wow, what a sight!.

Early in the morning, I got up to search out some lumber that I could fashion into mast supports. We would have to take the masts down to transit under the low bridges of the Erie Canal. I met two guys in the yard that had been up at the bar all night. They were not drunk, but they were tired. I asked about getting some wood, and one of them, who happened to be the owner's son, showed me a place under the docks where there was used lumber. I asked about hammer and big nails, and he came up with them too. These guys did not want money for what they gave to me. New Jersey generosity! We invited them aboard for some breakfast, which they needed. We sat in the cockpit while they ate. We asked if they would like to go below for a look around. One of them refused because of some kind of fear.

With our new crew, we motored up the Hudson River passed West Point and on to the entrance to the Erie Canal. The Erie Canal is 363 miles long. It lifts vessels up 556 feet to the level of Lake Erie at Buffalo NY. There are 35 locks on the canal to lift boats from 5 to 15 feet at a time. This was again a very special trip. At one place we tied up to get fuel, and asked if there was a grocery store nearby, thinking we could walk to it. The owner said no, the nearest store was about 5 miles away. He generously gave us the keys to his car, and let us use it at no charge. The milk of human kindness!

In Buffalo, we put the masts back up, changed crews, and headed for home. I remember entering Lake Superior through the Sioux locks. It was cold, and fogged in. As we left the lock, we heard a loud low horn of an approaching freighter. We couldn't see it, but I knew that if we got in its way, we were toast. I located a channel buoy, and made very tight circles around it. Finally the bow of the ship showed up through the fog. What a relief, we let the freighter pass, and then continued out into the lake.

Lake Superior is very cold, large, and deep. Its temperatures in summer get to about 50 degrees, and much colder in the winter. We were sailing across Lake Superior in May. We had the cockpit enclosure fully up, and we ran the reverse cycle air conditioner/heaters. They were powered by 110 Volt from our 8KW generator. Those little heat pumps took in sea water, extracted some heat from it, and returned the slightly colder water back into the lake.

We arrived at Lakehead Marina on Minnesota Point, Duluth MN after 22 days in transit from South Carolina. The trip was a very rich experience, and it gave us new respect for Vonnie T.

Voyage Preparation

It was May 1986. We planned to start our big trip in July 1987, right after Sara's graduation from high school, and her birthday.

I bought a couple of hundred dollars worth of charts, and would often spread them out on the living room floor planning and dreaming. I installed an inner stay on the foredeck of Vonnie T, and bought two new sails. The idea was that if wind conditions were high we could use a smaller jib on that inner stay, and if the wind was really strong, we could fly a storm jib. The main jib was a 170% Genoa Jib on roller furling. It was to be our power sail.

We refinished all of the topsides teak.

I bought a sextant, and taught myself celestial navigation.

Lavonne studied for her Ham Radio License. After a lot of work learning the Morse code, she passed the test with the help of some very kind Ham's, Ralph Andreas and Warren Koppy. These two men, and others that LaVonne met on the air would later talk to us on the Ham Radio at various places around the world. It was also possible for the Ham's that Lavonne contacted to do phone patches for us to our family. The contact was made Radio to Radio, and on the US end, the Ham had a device that could hook into the US phone system. All of this was done very kindly, and at no cost. Imagine being 1000 miles at sea, and talking to sons and daughters in Minnesota!

The Radios required installation, and required very good antennae, and ground systems. I struggled to do it just right.

We joined the Seven Seas Cruising Association (SSCA). They printed a monthly booklet about what members were doing and gave tips to members that were going their way. We identified ourselves as being members by flying the SSCA Burgee.

LaVonne bought a second hand 'Viking' heavy-duty sewing machine. She would use that machine to do sail repair. She also used it to make courtesy flags for any country that we entered. A courtesy flag is flown below the starboard spreader, and is a miniature of the flag of the Country. As we would enter a new Country, we would hoist a yellow 'quarantine' flag on the starboard spreader. After we cleared customs and immigration, we were allowed to replace the 'Q' flag with the country's Courtesy flag.

We talked to our Doctor about what we might need in the way of medical supplies. Well, that is a very tough and large topic. We ended up with a lot of stuff that we thankfully didn't need to use. We got yellow fever, typhoid and some other shots.

I visited our local Perkins dealer, as we had a 4 cylinder 85 hp Perkins diesel engine. I wanted to have a supply of spare parts that I could use to repair the engine.

We arranged with Brother Mike to be our mail drop. Any mail that we would get was directed to his address. At certain points along the way, we would have him forward a batch of mail to us (eliminating magazines and commercial stuff). The mail was sent usually a week or more from the time we expected to be in a particular place. It was addressed to S/Y Vonnie T, % xyz post office, City, Country. It was marked 'hold for pickup'. The large manila envelope had a yellow tape around one end. That tape would later help us pick it out at the post office when we had difficulty talking (because of language) to the postal people.

We used our VISA card as a bank account. We kept track of what we had drawn on that card, and when it got a bit low, we sent VISA a check for several thousand of dollars out of our home bank. We used that card all over the world to get cash in the currency of the country.

We sold our house, quit our jobs, and installed Sara into a dorm at Mankato State University. We had a big Bon Voyage barbecue on the lawn near the boat. We had about 100 guests. We gave hugs to family and friends, and already started to regret that we would not see much of them in the coming months, and perhaps years. It was a joyful and tearful time.

Off to Sea

I was 50 years old when we pulled out of our slip at Lakehead Marina, and headed for the ship canal. The mainland part of Duluth is separated from Minnesota Point by the ship canal. The canal is about 600 yards long, 40 feet deep, and walled with concrete. It is an official waterway, and as such it is required to have a bridge that will permit automotive traffic. That traffic is to take second priority to any ship traffic according to our maritime laws. The bridge that spans the ship canal is an Aerial Bridge that moves up and down some 135 feet to accommodate large ships. A bridge captain on duty controls it 24 hours a day.

As we approached the bridge for the last time, we gave a horn signal to indicate we wished to transit under the bridge. Our mast stood 65 feet above the water and would only require a partial opening. I talked to the bridge captain on the VHF radio. I explained that this would be our last transit of the ship canal, and gave a brief explanation of our plans. The bridge rose, and we sailed out into Lake Superior.

I made the following entry into our Ships Log:

46°North Latitude, 92°West Longitude, July 21, 1987

"MAN CANNOT DISCOVER NEW HORIZONS WITHOUT HAVING THE COURAGE TO LOSE SIGHT OF THE SHORE".

Duluth Aerial Bridge

We were only a few miles out in the lake when we got a call over the VHF.

"VONNIE-T, VONNIE-T, do you read? If you are still in radio range, this is the Duluth Bridge Captain. A tornado has been sighted. Return to port!"

Our trip of a lifetime was threatened to end nearly before it got started. We were faced with making a major decision, and doing it fast! The sky was still sunny and clear overhead but a black horizon was rolling toward us.   We considered turning back and trying to outrun the tornado as the Bridge Captain suggested, but we feared the storm might catch up with us just as we were negotiating the ship channel. Worse yet, the storm could catch us after arriving in the harbor where we'd be near rocks, docks and other boats with almost no room to maneuver. If that happened, our lives would probably be saved, but the likelihood of doing damage to our boat and others around us was highly probable.

One alternative was to stay out on the lake and take our chances. What should we do? That was the question.
The answer came quickly. We would stay our on the lake, stow the sails, and motor into the wind and waves.

The sky got dark, and the winds increased to perhaps 50 knots. The winds veered around, coming out of the north west, and then from the north, and then from the north east. We took a pretty good beating. I was worried that our nice cockpit enclosure wouldn't take the beating. It did.

The whole thing lasted less that 20 minutes, but it seemed like hours. The sun came out again, the skies cleared, we got a nice 10 knot wind, and we headed for the ending of our first leg, Bayfield WI.

We sailed across Lake Superior and through the Keewanaw peninsula cut where we had been several times before. Our route was across Whitefish bay, and over the wreck of the Edmund Fitzgerald. The Fitzgerald was a big ore boat that perished in a storm in 1975 taking all 29 of its crew to the cold depths of Lake Superior. Gordon Lightfoot created a song as a tribute to those men.

We transited the Sioux locks between Superior and Huron, and entered Huron via the Saint Marys River. A strange thing happened in that river. We were south bound, and we passed a small sailboat named Cathleen that was north bound. About a half hour after passing the Cathleen, we heard an SOS call on the VHF. It was from the Cathleen to the Coast Guard. The captain of the Cathleen was reporting a Man-over-board. The Coast Guard responded with giving the location of the incident to anyone near, and asking for assistance. They asked the Captain what they could do for him. The reply was "Nothing for me, I have been searching for my lost crewman. He is dead".

That chilling message made us redouble all of our preparations for safety at sea.

We transited Lake Huron and Lake Erie and prepared to transit the eight locks of the Welland Canal that separate Lake Erie from Lake Ontario. Those locks drop ships 326 feet. There is a lot of ship traffic through the locks. One of their rules was that each small boat must have a pilot, and two line handlers. Those line handlers would tend to the fenders that kept the boat from damaging its sides against the rough concrete walls of the locks. They were also to see that the boat was properly tied up in each lock for the lowering to the next level.

We were short one line handler. We met another boat that had just transited the canal northbound. They had a line handler who was headed back to Toronto on Lake Ontario. We made a deal with him to help us. He was a character. His name was Scott, and he was 6' 7" tall. He worked in a Madrigal Dinner theater in Toronto. He was a singer/actor. When we entered one of the locks for lowering to the next level, we noted a large observation platform where visitors could observe the lock operations. After the lock gates closed, and the water was slowly released the Vonnie T descended while a hundred eyes watched. Our line handler could not resist performing for that crowd. Scott burst forth with song in his deep and penetrating voice. His powerful voice echoed off of the huge walls of the lock. The crowd on the platform clapped, he bristled with pride, and we thoroughly enjoyed the whole scene.

Scott had to be back in Toronto for work at about 7 PM. Through delays in the lock operations, it was clear that we would not be able to get him there in time. LaVonne put a call out on her Ham Radio for help. A kindly Ham responded, and arranged for a small charter airplane to take our man back to Toronto. The Ham met us near one of the locks, and would drive our man to the airstrip. I shoved a hundred dollar bill into Scott's hand and bid him farewell. Those Ham's are wonderful.

Toronto was a wonderful place to visit, as were Montreal and Quebec City. Upon leaving Quebec City we encountered a lot of whales. It seems that a huge current flows westward from Newfoundland and through the Gulf of Saint Lawrence. That current brings a wealth of nutrients that the whales are keen to feed on. We sighted about 40 whales that day as we crossed the gulf. The best was a white Beluga that surfaced only about 6 feet from Vonnie T. As he came up he pushed ahead of him a lot of water that appeared like a big calm bubble. We watched him blow, and could see his eyes, and the wrinkles in his skin. He looked at us briefly, and slowly dove to the depths to continue his feast.

To digress, whales do not want to mess with boats. In fact they have such keen senses, that they can easily avoid collision with a boat. The only time that a collision could happen is if the whale was sleeping on the surface, and the boat struck it.

We tied up at Perce' Quebec. It was a small town with a marvelous restaurant that used a wood fired brick oven. We met the owner, and shared sea stories with him. He was a diver, and lobster lover. In season, he would catch lobsters, and put them in large cages that he had anchored to the bottom in 15 feet of water. Apparently, the lobsters would breed in that environment. He took restaurant leftovers and scraps out to the cages daily. He featured lobster on his menu all year round. Lobsters are bottom feeders, and live off of the dead critters on the ocean floor. Why do they taste so good? Was there a symbiotic relationship between the lobsters, and the restaurant?

When we left Perce', I got careless with one of our 5/8th inch Dacron lines that controlled the Genoa sail. The wind was against us, and we motor sailed close to the wind. Somehow that line got free in the water, and found its way to the propeller shaft. It got wrapped around the shaft. I quickly put the transmission in neutral to minimize the damage. Now what do we do? We had a few hours to go to get to our next anchorage. We wouldn't be able to make it with no engine, and a crippled head sail. So we turned around and headed back to Perce'. LaVonne got on the VHF to try to locate a diver that could help us in that cold water. The reply came from the restaurant owner who promised to send out a diver. As we approached Perce' again, a small boat came to us with a diver in a wet suit. He motioned to us to take the sails down. He then dove to the propeller shaft, and unwound the line. Wow, were we grateful. We were once again a functioning sailboat with a functioning motor. We tied up to the pier, and went into the restaurant. I tried to pay for the rescue effort. It seemed to me that that should be worth at least $100 or more. The owner would not accept money from me. He explained that when a mariner was in trouble, it was the responsibility of other mariners to help. I would remember that lesson, and later reciprocate to other mariners in need.

We were sailing across the North Humberland Straits during lobster season. At first we saw floats that were tied to lobster traps, and didn't know what they were. We avoided them, and tried to get around them without success. Then we saw lobster boats approaching each float, reeling it in, and winching the trap from the depths. This repeated over and over. We headed for the New Brunswick shore in hopes of anchoring in sand. One of the lobster boats approached us and began a conversation about where we were headed. I told him, and he said "No No". His native language was French, but he spoke some English. He invited us to tie up in the lobster boat marina. He directed us to the gas dock, and helped us tie up. As we entered that small harbor with perhaps 70 or 80 small lobster boats, I felt like a lobster entering a trap. There was precious little room for us to turn around. Once tied up, we became the spectacle of the village. I surmised that they never had a Yacht there before. Many, many of the local people came out for a look-see. They stayed on the bulkhead, but curiously peered into each porthole. Then the lobsterman who directed us into his harbor appeared. He gave us a cardboard box with 8 or 10 lobsters. Wow, what hospitality. He explained to us that these lobsters were a bit too small to take to market. There was some size limit in the marketing process. We had a feast that evening.

The next morning, I arose before sun up only to find the harbor was empty. All of those boats were out to sea doing lobstering.

Tides

As the moon passes around the earth some 27 times per month, the gravitational attraction between the moon and the earth creates tides. The moon pulls the water westerly with it. A huge part of the ocean is now in motion towards land. As the bulge approaches land, the water depth increases, and it seeks any inlet in which to flow. The tide at Quebec City is about 10 feet high. At Quebec City, the tide is neutralized by a lock. The inner harbor there is called Basin Louie. At high tide the lock gates are left open because the level of Basin Louie is kept at high tide level. As the tide goes down, the gates close and one must use the locks to pass into the basin. That basin is a very nice place, and within walking distance to a fine city. Tides can be a boatman's friend if he uses the tidal flow to assist his transit. At Quebec City, the tidal current can be as much as 7 knots. If the boatman tries to go against the tidal flow, he faces an uphill battle.

We journeyed to Prince Edward Island, and tied up in a little marina at the town of Summerside. We met some very nice people there, and stayed a few days. We planned to leave early one morning. The people there explained to us how the tidal flow affected the waters around PEI. There are two high tides each day, and two lows. The time between a high and the next low is about 6 hours. PEI lies roughly east to west. As the tidal flow comes in from the sea, it can reach speeds of 2 to 3 knots. At the end of each tidal flow cycle, the flow reverses itself.

We wanted to go from Summerside to Wood Island ferry harbor. The trip should take only about 6 hours. We would leave Summerside at the time that the tidal current was switching to outflow. We would get a boost from the tidal flow, and not have to go against it. Well, we spent way too long saying adieu to our new friends, and doing a little provisioning. We missed the optimal time by about 2 hours. In retrospect, we should have stayed an extra day, and left at the proper time. I thought, no problem, we had a good engine, and could buck the current if necessary. WRONG1

The day at sea started out nicely. Sunshine and a light breeze greeted us. Sometime about midday, things changed. Clouds darkened the sky, and a brisk wind began to build. Soon we had 20 knot winds on the bow, and needed to motor sail. The seas were rolling at about three to four feet. Our boat speed dropped, and I knew the tidal current would switch soon. We were still a long way from Wood Island. Then a light rain started. It was cold even with our cockpit enclosure all buttoned up.

It was well after dark when we approached the entrance to the ferry landing at Wood Island. We were happy to see that big ferry boat tucked comfortably inside the harbor. We knew that we would soon be inside too. Wrong again! We looked for navigational lights to guide us into the harbor. All we could find was a single red light. The rule at sea is 'red right returning'. We approached the harbor entrance such that the red light would pass on our starboard side. As we got closer to the rocky breakwater, it was evident that we were headed for a collision. We circled back, and had a think. Surely that red light was to guide us. We approached the entrance again, holding the red light on our starboard. The same thing happened. Now, what to do? We knew the entrance was there, but because of the wind and rain we could not find it even with our powerful searchlight. LaVonne called the Canadian Coast Guard on the radio seeking help. She told them where we were, and asked about navigational lights. The guy on the other end of the radio connection pretty much blew us away. He told us to check our chart. Our chart showed the entrance, and I assumed that it would be well marked with red and green lights.

We were now cold, wet, and afraid. We had been doing circles for at least a half an hour. Then we noticed a car with its headlights on the far side of the harbor. The car was positioned to allow the headlights to shine directly down the channel. We aligned Vonnie T with the car lights, and proceeded very cautiously in the direction of the breakwater. Sure enough the channel entrance appeared, and we motored in. The person who positioned the car lights for us was a young son of the lighthouse keeper. He and the ferry night watchman helped tie us up along the harbors bulkhead.

We were safe at last. I surveyed the light situation again. In fact there were no navigational lights there. The red light that we saw had something to do with the ferry dockage.

The young man that saw our dilemma and responded so beautifully came aboard. We shared some food, and told each other our stories. He was so interested in the sailing adventure that we had planned that he asked us to adopt him, and take him along.

As it happened, the next morning the sun came out, and the winds had died. We walked around the harbor, and the adjoining wetlands. It happened to be low tide. The sea bed was exposed, and we were pleased to find an abundance of clams fresh and ready for the picking. Life was good again.

In retrospect, we could have died that night had the boat foundered on that rocky shore. The whole problem was caused by my poor planning about the timing of leaving Summerside.

Nova Scotia

We are about to indulge in the best meal that I have ever eaten. We approached the little town of Canso on the northern tip of Nova Scotia looking for the fuel dock. After taking on a fill of diesel fuel, I walked around, as I often did, to have a look see. I saw a fishing boat tied up to the big cement pier. It was off loading big baskets. Upon further inspection I saw that those baskets were teeming with live queen crab. The catch was obviously fresh since the crab legs were all wriggling about. I watched as many baskets were off loaded, and received by the market guy on the pier. I thought about dinner. I asked the market guy if we could buy a couple of the crabs. He said sure, and for a few bucks we had crab. Now, what to do with them? I asked the market guy how to clean them. He said it was easy. He grabbed a crab with three legs in each hand. He first struck the crab on the edge of the cement table in the center of the pier. Off came the top shell! Next, he whacked the table edge with the crab, and the legs separated. He swooshed them in some water and handed them to me. Yes, it was easy. We thanked him, and headed out into the sound.

We had a big lobster pot on board. We filled that pot with water, and set it to boil as we motored across the sound in the afternoon sunshine. We had to take turns eating that very fresh crab, and steering Vonnie T. No finer food can man eat!

Nova Scotia has a very unusual coast line. It is serrated. Spits of land go out into the sea a few miles. Between those spits are channels that lead to small towns or marine terminals, or sometimes nothing. There may be a hundred of these spits. Thus for navigators like us, we would have go into the waterway some few miles just to get to a place we could tie up. Then on the way out, we would have to reverse that few mile trip. A lot of time was consumed entering and exiting.

In one of those canals was a fishing pier where we wanted to tie up in for the night. There were two commercial fishing boats already tied up on the inside of the pier. When the fisherman learned that we wanted to tie up for the night, they moved one of their boats to the outside or windward side of the pier. They did that so we would have a more comfortable night. That was very kind. People continually amaze us with kind acts.

One of the fishing boats had an 8 foot shark lying on deck dead. It turns out that those were net fishermen, and when they retrieved their nets, they sometimes attracted sharks who wanted some of their catch. Their practice was to pull the shark aboard with a gaff hook, and kill it. They could bring the sharks to market, but get only a minimal price for them. This particular shark was diseased they said, and not suitable for market.

We got into conversation with the fishermen, and the subject turned to seasickness. One of the fishermen told us he gets seasick every time he goes to sea. We commiserated with him. He said that is what I do. He pointed to a yellow house a ways away, and said his wife and kids live there. He was able to have that house because of his income from fishing.

Sailing down the shore on the way to Halifax, I was trolling with a CD 14 Rapala bait. Now we had seen seagulls many many times. We had little use for them as they left their droppings everywhere, and made an awful racket as they tried to speak. I didn't realize how dumb they were. A seagull picked my Rapala out of the water thinking it was good for him to eat. Well, he got hooked, and flew straight up. My reel sang as he took out line. When I saw what was happening, I took the rod and reel in hand, and tightened the drag so the bird wouldn't take all of my line. The bird went back to sit on the water. Then it happened. Another dumb seagull swooped down and tried to steal the Rapala from the first one. I reeled in as much line as I could and then cut it.

Halifax is a wonderful city. It is located on an inner harbor that is some 10 miles in from the sea. The harbor is very large. As a matter of fact, during World War II, fleets of merchant ships would mass up there, They formed up into convoys to cross the ocean bringing supplies to the British who were deeply into the War.

As we approached the big pier in front of the maritime museum, we noticed a man there waiting for us. He helped us tie up, and explained that he had been listening to LaVonne on the Maritime Ham net every morning. He had followed our progress for about a month. He never spoke on the net, but he wanted to meet us, and do what he could to help us. That was very much welcomed. He had a car, and took us to the Laundromat where we could freshen up our clothes. Now, imagine that kind man going out of his way to help total strangers. Wow.

We also noticed Halifax hospitality in the streets. If a pedestrian is approaching the street looking like he wanted to cross it, the cars would stop, and let him pass even though there were no traffic rules that required it.

The Eastern Seaboard of the United States.

Dear reader: I have a thousand sea stories or maybe two thousand. We had some delightful times on the eastern seaboard. Instead of going through a travel log about those times, I will only list the memorable places, and suggest that you check them out when you can. These places are rich in American history.

Provincetown MA
Boston MA
Newport RI
New York NY
Chesapeake Bay VA
Norfolk VA
Beaufort SC
Charleston SC
Savanna GA
Saint Augustine FL
Fort Lauderdale FL

The Bahamas

Who ever heard of 'No Name Harbor'? Well it is a little south of Miami. It is a place where small boats amass getting ready for the short passage to Bimini, Bahamas.
We anchored there getting ready for the passage.

We needed to plan for the current running outside, and to the north. The Gulf Stream flows at about 2 knots northerly. If we struck a course straight for our port in Bimini, the current would push us off course. We would probably make 5 knots. If the crossing was 30 nautical miles, then it would take us about six hours to cross. That gave the Gulf Stream six hours to push us northward.

Lets assume that we setout for only one hour. The one hour would allow forward progress along the hypotenuse of a right triangle of 5 nautical miles. Meanwhile the current would set us northward 2 nautical miles. The angle at which we must sail is that angle which has a sine of 2/5 or .4. The trig tables show us that the correct angle is 24 degrees.

The Gulf Stream is the only open ocean current that we would have to consider in our navigation. There are other open ocean currents, but they are usually less than $\frac{1}{2}$ knot speed, and would have only minimal effect on our travels.

Back in No Name Harbor we were about to experience a phenomena that I had read about. It is called the 'Herding Instinct'. The name is derived from the activity of cattle. When one cow decides to meander to the far side of the pasture, the others will follow. When we heard of this phenomenon, we vowed it wouldn't happen to us. We would plan our voyages based on what was right for the Vonnie T. If others would alter their departure times to follow the 'herd', it wouldn't be us.

Sure enough it happened. We had calculated that the passage to Bimini would take us about six hours. We would have to clear Customs and Immigration when we arrived. We always tried to plan voyages so that we would arrive in the new port or anchorage in good daylight. We figured that if we left at 3 AM, we would arrive at Bimini at 9AM. If we spent two or three hours doing the customs thing, we would exit Bimini about noon, and still have plenty of time to find a good anchorage for the night. Well, guess what happened about 1 AM. The first boat left No Name, and headed for Bimini. Soon after the second boat left and so on. By 3AM we were the only boat at No Name. What a good deal, all of those other (about 15) boats were now out of our way, and most would have cleared customs before we got there. The 'Herding Instinct' had worked to our advantage.

We checked in at Bimini, Bahamas, and proceeded to find an anchorage. The Bahamas are a group of perhaps a thousand islands. They are mostly pristine. They have a lot of very nice sandy bottoms, and very clear water. We learned to 'read' the water. You can look across the water ahead, and tell the depth from the color of the water. The deepest blue was very deep, as the color turned light blue, and even sandy brown, watch out. Grounding was possible, and that did happen to us a couple of times. We had a depth sounder, but by the time it read 7 feet or less, it was too late.

We learned to spear fish with a 'Hawaiian Sling', the only legal spear fishing device in the Bahamas. The fishing was very good. We had a steady diet of fresh Grouper, and Yellowtail, and whatever else got too close to us.

In Nassau, LaVonne had to go to the Communications people, and apply for a reciprocal Ham Radio License. The license was pretty much on hand for the asking, but to use the Ham Radio in Bahamian waters without it was illegal. Lavonne could now enter the 'Caribbean Net' every morning and be in touch with other boaters many miles away. She would follow the same procedure in all of the countries we were to visit.

Our navigation up until this time was mostly dead reckoning, using compass and charts. Now we started to use our Loran C. That system made contact with two shore stations, and used triangulation to pinpoint our location. Loran C is good up to about 100 miles off of the US coastline. It is now obsolete. In Nassau, we bought our first Sat Nav. The Sat Nav communicated with satellites as they passed overhead. With the Sat Nav, we could get a 'fix' on our position each time a satellite passed. We got a fix about once every two or three hours. Since we were traveling at about 6 knots, it gave us location every 15 to 20 nautical miles. The Sat Nav would work any place in the world.

One of our anchorages was at Norman Island, Bahamas. There was a nice resort there, and an airstrip. As it turned out, the authorities there, in conjunction with the US drug people, had closed down the resort a few months earlier. It had been used to fly drugs into the US. It was in a sad state of disrepair. There were no people there. We went into a couple of the rental units. It was a very eerie feeling to know that this place was once a thriving resort feeding the drug business. I guess that whoever ran the place was now in jail.

It was now Christmas Holiday time in 1987. We went to Nassau to pick up family. Clark, Sara, Andy, and Andy's friend Ken were there to be picked up at the airport. Wow was that ever nice to see them again. We sailed out into the more remote Bahamian Islands, and did a lot of snorkeling in that warm water.

Diving

I had been a pretty good swimmer all of my life. Now, on our big voyage, I would become a good free diver, and SCUBA diver. We often went on snorkeling expeditions to sightsee the wonders of the ocean bottom. There were reefs with all sorts of colorful coral, and a myriad of large and small fish. The water temperature was in the 80 to 85 degree range, very comfortable. Equipped with a mask, snorkel, swim fins, and spear gun, I often spent two or three hours a day in the water. I would do a surface dive, and then swim down to the bottom. I surprised myself one day, as I dove looking for fish, I noticed the dinghy anchor bobbing on the sandy bottom. It had 35 feet of rope on it and a bit of chain. I was free diving on only the breath I took at the surface in 35 feet of water. I found that as I looked for fish, I could always go a bit deeper, or stay down a bit longer. I remember one day as I returned to the surface that my legs were feeling tingly and numb. I had stayed down too long, and was short on oxygen. The next thing would have been a black out. I crawled back into the dingy, and had a talk with myself. In the future if I felt the need for air, I would go to the surface and get it, and not flirt with disaster.

My favorite catch was a Grouper of about 3 or 4 pounds size.

 I also brought back Conch, Squid, Clams, Oysters, Lobster and anything that looked like good eating. We would then look in our fish book to learn what kind if fish it was, and whether or not it was good to eat. Pretty much all of the fish were edible. The exception in the Caribbean was Barracuda. Barracuda feed off smaller reef fish, and get a build up of Ciguatera from the coral that smaller fish eat. Ciguatera poisoning in humans causes a reversal of heat sensation. Cold things feel warm, and warm things feel cold. We met some other cruisers who had a touch of the problem. They told us that they pretty much stayed out of the water, as it felt too warm to them. The condition lasted a few weeks.

Barracuda and Sharks have a bad reputation as people killers. I think the whole thing started as a result of the movie Jaws. Barracuda hunt in pretty big schools of maybe 50 or 60 fish. They split up when they are over a reef where their quarry is trying to hide. I have been in the water with pretty big Barracuda, maybe 40 pounds or so. They were definitely not interested in me. Likewise we have been in the water with Sharks perhaps as large as 6 or 7 feet in length. Those Sharks are also not interested in humans. They feed off of smaller fishes, Sea Lions, and Seals. You may have seen photos of a Shark bite on the torso or leg of a human. Those grizzly photos show a horseshoe shaped series of skin punctures where the Shark supposedly tried to take a bite. In reality, the Shark mistook the human for something else, and when he realized he made a mistake, he let go. He could easily have taken a huge bite.

One day when we were out doing our thing, LaVonne snorkeled on the surface as I hunted. I shot a nice Grouper in about 12 feet of water. The Grouper in his attempt to escape took himself, and my spear through a hole in the coral. To retrieve my spear, I dove and tried to get the Grouper off of the spear, hoping to then slide my spear back to where it came from. A couple of dives, and I still didn't free my spear, but I noticed a Shark about 5 foot long swimming back and forth, He sensed that my wriggling Grouper was injured, and was there for the taking. Now, if I had my hand on the Grouper at the instant that the Shark attacked, I might have lost my hand. The Shark wouldn't be looking for my hand. My hand would have been in just in the wrong place at the wrong time. So, what to do? I wanted that spear back, and hopefully the Grouper too. I watched the Shark circling, and asked Lavonne to get between the Shark, and me and violently splash the surface. She did, and it scared the Shark away. I retrieved my spear, and the world was OK again.

Dominican Republic

We had Andy on board now, and exited the Turks and Caicos for the Dominican Republic at Puerto Plata. It was a nice crossing, and we tied up at the customs dock there. I remember the customs guy asking for a 'propino'. Normally that meant tip, in this case it meant bribe. Not having any of that, I pretended to not understand him, and walked away. He could have made big trouble for us, but not today. The remarkable thing about Puerto Plata was the bicycle traffic. They used bicycles to transport everything. It was common to see a guy on a bike with a plastic bag full of bread. That bag could easily be five feet in diameter.

We also met a young man about 13 who told us about lobster fishing. He was in fear of losing his life. The lobster rig was a boat of maybe 80 feet in length. It carried a bunch (12 to 15) of small boats equipped with oars and a small air compressor, and 200 feet of rubber hose leading to a secondary SCUBA regulator. Their fishing 'crew' was 12 to 15 young boys that needed money. They went to a likely lobster spot and set the small boats over the side with the SCUBA set up and a young Dominican. The boys would then start the compressor, and go over the side wearing a dive mask, and the regulator. Now, these boys had no training about SCUBA, and the dangers of going too deep or staying at depth too long. Apparently the older boys told the younger ones what to do, and not to do. Once at depth the boys would gather lobster. They were paid by how many lobster they brought back. They were often tempted to go too deep, or to stay down too long. Each fishing trip lasted a few days. Our young man told us that sometimes when they returned, they brought back the corpse of a young diver.

He asked us to take him with us to Puerto Rico. We wanted to help him, but knew the way the Puerto Rican's handled stow-a-way Dominicans. There were so many Dominicans that wanted to go to Puerto Rico where the living was easier that The Puerto Rican government would nab the intruder, and put him on an airplane headed home. The Dominican government would have to pay the airfare. What to do? We gave him enough money so that he wouldn't have to go out on the next lobstering trip, and advised him to get some education to allow him to get a decent job.

Next stop was Samana, Dominican Republic. We set out one afternoon, knowing it would be an all night voyage to Samana. Andy was sleeping and LaVonne was at the helm. It was time for me to relieve her. A pretty good sea was running, maybe 5 feet. Before going topsides, I made myself a pot of coffee. As I was pouring the coffee into a cup (one hand on the pot and one hand on the cup) a rogue wave hit us. Instead of dropping the coffee, I held on. I fell backwards against the chart table chair, striking my lower back. I felt an immediate intense pain and fell on the cabin sole near the ladder. It hurt so bad that I nearly passed out. I could not move with out having a stabbing pain in my lower back. LaVonne immediately woke Andy to take over the helm. She came to help me. She said I turned ashen color, and sweated profusely. What to do? She found some pain reliever in our first aid kit, and gave it to me in triple dose. It worked, and I dozed off, still lying on the sole. Andy and LaVonne guided Vonnie T into the Samana harbor and anchored. By that time, I was able to crawl back to the settee, and get my bones up on the mattress. The pain was still there any time I moved, but was OK if I just lay still.

Now, Andy and LaVonne went ashore to check us into customs and immigration. At customs, LaVonne presented our boat 'papers'. She was told to take a seat. It turns out that they did not want to deal with a woman, and she was told to bring her 'Captain' to the office. This was all in Spanish of course. Eventually, another boater came into the office, for check in. LaVonne told him what had happened. He was pretty good with Spanish, and relayed that to the customs guy. Eventually, the customs guy agreed to ride out in our dinghy to do the check in on our boat. When he saw me in my sorry state, his manner changed. Quickly we signed the papers, and he was off.

Imagine the arrogance of the customs guy to not deal with a woman!

Later, LaVonne contacted a Ham friend who happened to also be a boater, and who was a doctor. She described what had happened to the doctor over the radio. The doctor quickly diagnosed the injury as a ruptured kidney. He said the best thing was to stay down for a few days. He said, "Do not go in for medical help in Samana". They might want to operate, and then what? The next day, I was feeling some better, and could move about if I did so carefully. A few days later, I was able to go ashore. A week later, there was still pain if I stepped off of a curb onto my heel. But, I was feeling good enough that we took a bus trip to the capital 'Santa Domingo'. One of Columbus' tombs is located there. There is also one in Cuba, and another in Seville Spain. We saw two of the three, and were told by each that they had the 'real corpse'.

The next stops were Puerto Rico, Virgin Islands, St Martin, and the rest of the Grenadines. All are wonderful places, and all have a bunch of 'sea stories'.

Our best Lobster fishing happened in Los Aves Barlovento in Venezuelan waters about 80 miles north of mainland South America. One of our buddy boaters knew how to get into an area with a lot of small reefs. Actually there were three of us Cruisers that went in there. There were a lot of lobsters. Andy was with us. On the first day we were snorkeling looking for Lobster. There was a small coral formation shaped like a volcano in about 12 feet of water. As we peered into the opening, we spied a very nice 10 pound Lobster. That was our first of many lobster kills. We ate Lobster for breakfast, lunch and dinner for a couple of days. Then, like an overdose of chocolate bars, the rich meat lost its appeal. We switched to other fish.

Another interesting thing happened in that anchorage. One day as we were snorkeling, we noticed a dinghy and motor had washed up on the reef. We towed it back and tied it to Vonnie T. It was in good shape. We put out a call on the VHF, and Ham radios looking for the owner. After 3 or 4 days of no response, we counted it as lost at sea, and claimed it. Now, how do you split up a dinghy among three boats? Vonnie T had a good 10 foot Avon with a fiberglass bottom, but with a used 10 HP motor. We determined that Vonnie T would get the nearly new Yamaha 15 HP motor, The used 10 HP motor was put on the found dingy, and given to one of us that needed a dinghy, and the receiver of the found dingy would give the third boat $250. It all worked out for the best. We would rather have returned the found dingy to its owner, but that was not to be.

I had a very special view of a female lobster in the birthing process. She was in about 15 feet of water. Now, female spiny Lobsters have small pincers on their rear legs, males do not. My lady Lobster was busy taking eggs out of her body opening. Previously, a male Lobster had deposited Semen on her carapace, a bony structure on her chest. Mama Lobster was dipping eggs in the semen, and planting them under her tail. Both rear pincers were very busy planting what was to be her new family. I understand that there are thousands of Lobsters planted in each birthing cycle. When those tiny Lobsters are later released into the water, they become Zooplankton. Much of that Zooplankton is eaten by fish in the food chain. In Mother Nature's carefully orchestrated plan, enough of the tiny Lobsters survive to develop into adults, and continue the species. I dove 15 or 20 times to observe the wonderful process. I wished that I had an underwater Video camera to record the event.

Ham Radio

We had two Ham Radios on board. The 20 meter was for long range communication from a few hundred to two or three thousand miles. LaVonne used the 20 meter radio to contact our family via phone patches with Hams in the U S. LaVonne's call sign was NOHWB. Each time a Ham called on the radio for a contact, he/she identified themselves by call sign. That was their method of knowing the other party had a legal Ham license. She used the 2 meter radio for short range (up to 200 miles) daily communication with other boating Hams, and land based Hams on Islands that we would sail to. The daily 2 meter net met every morning at 7 AM was called the Caribbean net in these waters. On it she would get information from boaters as to anchorages, Customs and Immigration, fishing, and a myriad of other things. She talked to land based Hams on pretty much every island in the Caribbean. The land based Hams gave us valuable insights about their island, and often would invite us to meet them and come to their homes.

LaVonne the Ham

One of the land based Hams Lavonne that met on the radio lived on the island nation of Saint Lucia. After anchoring in St. Lucia, we took the local bus the Castries, the capital city of St. Lucia. We met the local ham named Johnny and were warmly welcomed. We shared sea stories with him, and were given a private tour of the Castries. The most memorable thing on that tour was going past the fire station. Our Johnny explained to us that that new looking fire engine was a gift from the Canadian government. A new one in Canada had replaced this fire engine, but this used one was in very good shape. He was very proud of that fire engine. As we traveled, we found other governments that had given to poor islands.

Later, in Raiatea, French Polynesia, we would see a new breakwater, and commercial pier that the Japanese government had built for them.

We began to think about all of the U S foreign aid dollars that were spent each year. To my knowledge, that foreign aid is all in cash. The government of Israel gets about $5,000,000,000 per year. The government of Egypt gets about $4,000,000,000 and so on. Much of that foreign aid cash goes to buy military stuff. I would guess that a nice chunk of our Foreign aid is lost to corruption. Wouldn't it be better if we gave foreign countries tangible things like fire engines, and piers?

In Dominica, Lavonne had become Ham friends with a large female local, Martha J73MC. When we anchored there, I was greeted by a few punks who tried to extort money from me for the pleasure of anchoring in 'their' bay. I never bowed to petty extortion, and refused their help as I did the anchoring job myself. Later the punks came near Vonnie T, and looked menacing. LaVonne got on

151

her 2 meter radio, and contacted Martha. Martha was disgusted by the punk story, and called the local shore patrol to come to our aid. Within $\frac{1}{2}$ hour, a small Dominican patrol boat came out. The punks vanished. I chose to stay on the Vonnie T while LaVonne got a Dominican tour from her new friend, Martha.

Another Ham met us soon after we anchored in Trinidad-Tobago. His name was Mohammed Ali (just like the boxer) Y94BG. He immediately escorted us to his home, and wanted us to stay there. Again we were treated as good friends, and got an insiders view of the island..

LaVonne had been in ham radio contact for some time with Tony, J39CM, who lived on Grenada, and we looked forward to having an 'eye ball' with him. Tony had gone to England for his education and was a well-spoken, interesting man. When we arrived in Grenada, he met us and took us on a tour around his island. He showed us a quiet blue harbor called Quarantine Harbor, and we walked through the bustling downtown and market areas and breathed in the sweet fragrance from the impressive red flamboyant trees and other tropical trees and plants. We got to see bananas being packed for shipment and learned that each stalk had a blue plastic bag placed over it during its growing period to prevent bugs from leaving marks on the peel. We also toured the nutmeg factory. Grenada is the world's largest producer of nutmeg and mace. The factory, a dusty place with ground nutmeg hanging heavy in the air, would never meet OSHA air quality standards in the United States, but it was a fascinating place and I had never realized that two spices came from the same plant. Mace was from the red membrane that surrounded the actual nut of the nutmeg.
Tony made sure we saw all the positive aspects of Grenada before he told us the truth about the military coup of 1983.

"It's not at all the way your American Press reported it," he said. "Your President did the right thing, but the press wasn't told about it ahead of time. They got their noses out of joint because they hadn't been informed in advance, so they reported the whole thing badly. Your president had to keep it a secret because the attack needed the element of surprise in order to be successful," he said.

Tony said he knew the press reported that American troops were called in to free U.S. medical students, but it was something much more than that. "We had a tyrant and his thugs running the island, and they had plans to march right up through the whole Caribbean and take over one island at a time. Cuba was backing the whole project and provided both funding and manpower to build our big airstrip. They, in turn, got their funding from Russia. On the heels of the Cuban Missile, crisis they needed an airstrip large enough to accommodate military fighter planes. The thugs had everyone on the island fearing for their lives and they had our President locked up."

"One day, some grade school children with all the innocence of young children united and marched into the President's courtyard. They sang and chanted a request to have our President released. They were sweet children, in their little navy blue skirts and slacks and white shirts or blouses, innocent and harmless, just expressing themselves."

Tony could hardly go on with his story. Tears ran down his face. Finally he said, "They were fired upon and literally mowed down - slaughtered, all of them. It was terrible! There were piles of little bodies. They shoveled the children's bodies up with road grader equipment, tossed them into boats and dumped them in the ocean. Anyone who got in the thugs' way was killed."

"In the middle of the night I was dragged from my bed and placed in an underground dungeon." He showed us the dungeon, unlocked the steel door covering the steps and invited us to go down into it. We went down only a few steps, enough to see that it was damp, dark and filthy - and that's what it looked like with the trap door open. I can't imagine what it would have been like to be locked in it for days, as Tony was, with no idea of what was to happen next. He told us how many steps there were and where every blemish was in all four walls, the ceiling and the floor, but he couldn't bring himself to go down into it. Apparently, the reason he had been arrested was because he was a ham radio operator. The thugs who took power didn't want any news of what was happening to get to the outside world - not while they were disposing of children's bodies.

For three days the entire island was under house arrest. If anyone was seen out of their home they were shot on sight, and of course the families couldn't even acknowledge their missing children. Some families lost several children. Not only did they have to grieve their losses silently, but many also went without water and food for three days because most of the homes on the island didn't have refrigeration or running water. Typically the families would gather what they needed each day, one day at a time.

"On the third day I could hear bombing and a lot of gunfire while I was in the dungeon. Helicopters were overhead. Then, I heard ground fire. My fear turned to joy because I knew the Americans were coming by air to free us. Then everything got quiet for a while. Pretty soon I heard voices, and people opened the dungeon door. I walked up those steps not knowing what would happen next. Then I saw a garden hose. I walked over to it, took off my filthy pajamas and hosed myself down. I stood there naked, clean for the first time in days and alive. Later, my wife and I were able to put all the pieces together and we're very grateful to you Americans!" Tony and his wife invited us to dinner that evening at their home. We heard the story again from the wife's viewpoint. She told us she was frantic when she had been locked in her house, not knowing all that time if her husband was dead or alive. She saved pictures and newspaper clippings from the entire event. She kept saying, "It's not true, the way your American press reported it. If your troops hadn't come when they did, a lot more of us would be dead and many of the Caribbean islands would be under Cuban control by now. Your President did it just right. He did a surprise attack, and he stuck around just long enough to appoint temporary leadership, then he pulled the troops out."

LaVonne and I felt very proud to be Americans!

Learning about the military happenings on Grenada first hand from Tony and his wife gave us chills. We dug deep into our memory bank, trying to remember how the American press had portrayed the incident. That was the first time I ever even questioned the accuracy of how our Americans press reported world events. I had always naively believed that Americans were the most well-informed people and that the information we were given was accurate and completely unbiased. Yet, here was a situation we knew had not been reported objectively or accurately. Tony's words kept ringing in my ears, "the press got their noses out of joint because your President hadn't notified them before the invasion." Tony had a great deal of respect for our President and acknowledged that not only was the element of surprise important, but he also was pleased that the American military power got in, did what they needed to do, and then got out. The problem was that the whole thing was done and over with before the press even knew what had happened, so they took a critical position rather than a supportive one.

I could go on and on about the kindness of the Ham community. One last story is about a Ham that LaVonne met on the air. His name was Eric Fogg. Eric and his wife lived in a very nice apartment in Bogotá, Columbia. They invited us to come for a visit. They had originally lived in the Netherlands, and spoke good English. They had prospered in Columbia, and owned a 'Finca'. Finca is small farm in Spanish. When we got to the Finca, we found it was not so small, and in fact had a caretaker couple living in an adjoining building. They just treated us royally. Once again, we got a view of the Bogotá area from an insider's viewpoint.

South America

We had visited pretty much every Island Nation in the Caribbean. Now it was time to escape the hurricane season. Hurricanes are boating disasters. The season in the Atlantic is officially June 1 to November 31. However the probability of hurricanes occurring before August 15th, or after November 15th is very small. Our NOOA weather charts predict the percentage chances for any part of the Atlantic. The hurricane belt is north of 20 degrees north latitude. We chose to go to Cumana Venezuela for the season. Cumana lies at 11 degrees north latitude.

Now, dear reader, I will skip around to a selection of the numerous places we visited, and tell the unique stories that made our traveling life very rich. I will try and not make it a travel log.

Our total 'land time' in South America was about nine or ten months. It happened over two hurricane seasons, and a long wait in Ecuador for a permit to visit the Galapagos Islands.

South America is a place seldom visited by Americans. In our travels there we met only a few. We found South America to be pretty safe. We always approached people openly, and they reciprocated in kind. We were bothered with petty theft a few times, but they were pick pockets, and snatchers, and never came at us with a gun or a knife. We found that the cost of food and lodging to be only about half of what would be charged in the U S. With practice, we were able to communicate pretty good in Spanish. I think we had about 'vente porciento de las palabras' (Twenty percent of the words). If someone wanted to communicate with us, it always worked. Portuguese is the language of Brazil. It has some similarities to Spanish, and once again we could communicate. Much of our vocabulary centered on travel, hotels, food, and boat parts. We found our way around mostly by talking to other travelers, and by following our noses.

Venezuela is a pretty unique country. It has a population of about 30 million. It lies, in part, in the Andes Mountains. It has a full set of natural resources: iron ore, coal, crude oil, gold, diamonds, timber, rich soil, and lots of water. It has a dam on the Oronoco River that is capable of producing enough electricity to serve all of Venezuela's needs. It is a member of OPEC, and is the world's eighth largest exporter of crude oil. It has the world's largest crude oil reserves. Venezuela should be one of the leading countries in the world, but alas, it has had over a long period of time a government that has been corrupt, and inept. From my view, education has not been a priority.

Our first hurricane season was spent in friendly Cumana marina. We had heard about it from other cruisers. We were tied up in a nice slip that had a supply of water and electricity. There were about thirty other cruising boats there, all escaping the hurricanes. We developed a nice social life.

We all used credit cards as a source of cash. I remember it being a two-day process at the bank. We lined up at the proper window in the bank with other people from around the world. We all filled out the forms, and showed Visa cards, and passports. Then we were instructed to come back the next day for our cash. Let me tell you that $100 in dollars gets you a stack of Bolivars about two inches high.

Cumana had a large food market. It was in a building that was about half a city block in size. Inside were stalls of vendors selling any kind of food you could imagine. There were kids that provided a carrying service. For about $.25 the kid would follow you with a wheelbarrow, and collect anything you purchased. We had to be careful because anything we put in that wheelbarrow would have to be later transferred to boat bags, and transported via bus back to the marina. One could easily get more than one could carry. My favorite purchase was 'Lomito', or a whole beef tenderloin. MMMM good.

Most nights, a bunch of us would jump on the bus, and go downtown for dinner. My favorite food was Pollo Embrasso, or chicken cooked on a spit over a wood fire. Shopping around that town was a lot of fun. I remember getting two Spanish words mixed up. Derecho means straight ahead, Derecha means turn right. Now when you ask directions, and get the correct ones, but go straight ahead instead of turning right, you get lost. Not to worry, just ask again. One treat that I had never seen before was fresh orange juice on the street. The vendors had carts piled high with fresh oranges. For about a dime, they would slice an orange or two in half and squeeze the juice out into a plastic cup. Wow that was good.

We flew into Caracas, and took a bus from the airport into town. Along the bus route were 'Barios', or poor neighborhoods. They lined the steep hills on both sides of the route. There were no streets, only dirt paths. The housing consisted of handmade shacks made with plywood, corrugated tin sheet, and cardboard. These people are very poor. Downtown was a very different place. It was a modern city complete with reflection pools, tall buildings and a very nice subway. We rode the subway one day, and were approached by volunteers taking up a collection for UNICEF. I pulled out my wad of Bolivars, and put one or two into the basket. Unbeknownst to me someone was watching looking for an easy mark. We went down the escalator, and rode to a suburb. There we got off of the train, and went up the escalator, and near the top it happened. The man in front of me intentionally dropped a pile of papers at the top of the escalator. As he bent over to pick up the stuff, he caused a backup of people who bumped against one another. I immediately thought 'Scam', and reached in my pocket to check my money. It was gone. I looked in back of me and saw a couple of men, any one of which could have been the pickpocket. The jam-up cleared, and everyone calmly walked away. Wow, I thought that was clever. They had followed us the whole subway trip, and set up the little scam. I was out about $50, but in reflection, had respect for a smooth pair of thieves.

We took a two month bus/airplane trip to Brazil. What an exciting and enriching trip. We first flew to Santa Elena Venezuela on the Brazilian border. We used our new back pack/suitcases. There is a compartment on the back that holds the back straps. When it is zippered up it becomes a suitcase. It has soft sides, so that we could easily stow them in a small place on the boat.

In Santa Elena we got visas from the Brazilian Embassy. We boarded a bus for the eighty mile trip to Boa Vista Brazil. Now that was a trip. It was all on gravel roads, and over some precarious 'Puentes de Madera' (wooden bridges). At one bridge, we were all instructed to get off of the bus, and walk across. It was a pretty rough bridge, and we had to watch where we walked. Then the driver had his assistant walk in front of the bus as it crossed and point out the safe way.

The eighty mile trip took us about four hours. (Twenty miles per hour)

From Boa Vista we flew to Manaus. Manaus is on the Amazon River and is subject to extremes in water height. In the dry season, the water level is fifty feet lower than in the rainy season. From

Manaus, we flew to Belem on the Atlantic coast. The Amazon is huge. At Manaus during the rainy season it floods an area perhaps 100 miles wide. At Belem, It is so wide that there is an Island in the river that is larger than the nation of Switzerland. The tides are pretty big there. At low tide we could see many boats sitting on the seabed after the tide ebbed.

From there on, we took busses down the entire coast of Brazil. We planned each segment so that we had a reasonable ride time. At each bus depot in each city, we bought tickets for the next city. If we liked the city we extended, and used the ticket later. Some cities were only good for an overnight. We did this all in Portuguese, and sign language. Rio de Janeiro was a pretty special place. We rented a small apartment only one block from Epenema beach. It cost $35 per night (guess what the cost in a beach hotel might be). Oh yes, there were a few cock roaches there. We stayed in Rio for 10 days. There is not enough paper here to describe all of the wonders of Rio; The beaches, Sugar Loaf Mountain, The Christo, Tram rides into the barrios and etc. The people of Brazil are lovely. The Portuguese entered the country many years ago. They brought slaves from Africa to work the fields.. There also was a large indigenous population there. Over the years, these three groups inter married, and interbred. The resulting light brown skin is very nice.

A family in Porto Alegre Brazil greeted us. Years earlier, LaVonne had hired the sister of the man of the house. The whole family was forced to leave their native country of Persia (Iran) when the Muslims took over from the Shaw of Iran. It seems the Muslims under the Iatolla were intolerant of anyone not Muslim. The family was of the Bahai faith. For them it was either go away or die. The sister came to Minneapolis, and got the job with the

University of Minnesota's extension program called 'Expanded Food and Nutrition' which LaVonne directed in Hennepin County. LaVonne kept in contact with the sister who was grateful for being hired. When the sister heard of our trip, she wanted us to visit her brothers' family in Brazil, who had also left Persia under threat from the Muslims. We were very warmly welcomed. The family took us to a Brazilian Churrascaria, a very nice restaurant that specialized in roasted beef, lamb, and pork. Wow what a meal!

Just writing about this some twenty five years later, brings very warm feelings to me. We were adventurous, and sought out new things constantly.

After Brazil, we went back to the boat in Cumana.

We also traveled by plane and bus to Canima National park high in the Amazon jungle, Pico Bolivar at 15,000 feet altitude by cable car, and Angel falls. Angel falls is the highest waterfall in the world, dropping some 1000 meters. All were marvelous places to visit.

The following hurricane season we anchored near Puerto la Cruz Venezuela. On leaving Venezuela, we sailed northward in a pretty big following sea, perhaps 6 feet. Vonnie T had an exhaust system that mixed engine cooling water with engine exhaust. That was done in a small chamber over the engine. The mixture was piped out in an upward loop of three-inch hose. The top of the loop was well above the engine, and normally functioned to stop seawater from entering from the rear. When we got to our anchorage we took down the sails, and started the engine to begin the anchoring process. I would stand on the bow and look for a likely spot. I then signaled LaVonne to put the transmission in reverse to stop the boat. The anchor chain was then let out, and LaVonne backed Vonnie T slowly until I locked the chain, and caused the anchor to imbed in the sea bottom. We would then test that it was properly set by going at about 1/3rd throttle in reverse. If the anchor held, we were sure that it would hold in any blow that might later occur. After anchoring, we went through a routine that included washing down the salt watery decks and lifelines, setting up things below that we had stowed for the crossing, and checking the engine.

Well, the engine oil was milky looking. Seawater had entered through the exhaust, and was pushed up and over the loop in the exhaust line by repeated assault from the following sea. If we ran the engine for any length of time in that condition, we would ruin the engine. The lubricating qualities of the oil had been seriously compromised. I got out our little pump that was driven from a drill motor, and drained the engine. I put a little more oil in and started the engine briefly. Then, I drained that oil too. Now I refilled the engine with clean Rotella T SAE 30 motor oil. I knew something had be done to prevent that in the future, and determined to find a bronze 3 inch check valve to install in the exhaust line near the place where it passed through the hull.

Our next stops were at Bonaire and Curacao, Dutch Islands about 50 miles north of South America. They were, once again, very pleasant places to visit.

We wanted to go to Columbia. We had heard the warnings from the U S State department about the dangers of going to Columbia. In those days the drug business flourished in Columbia. We had also been in contact with some cruising friends that were also Hams. They were in Cartegna, Columbia, and reported that there were absolutely no problems. It turns out that unless one was rich, in politics, or in the drug business, there was no problem.

So, we set sail for Cartegna. We knew from the NOOA Pilot charts that we could expect high winds and seas. As the southeast trade winds blow toward the west, and as they pass the northern part of South America, they get funneled, and speeded up by the Andes Mountains. OK, we were prepared for heavy seas. We would batten down everything loose below, and give it our best. We flew only the staysail, and departed Curacao one afternoon. It would be an all night and into the next day passage to Cartegna. The winds picked up as expected, and in the early evening we had seas that were 14 to 16 feet high, and winds that blew at 40 or 50 knots. It was rough, and we were going down wind, and down wave.

Now if we kept Vonnie T going straight down the face of those big waves, she would enter the trough, and rise up the back of the next wave. We were going fast. If we let her get sideways with the wave, she would likely dig in a leeward rail, and probably overturn. Overturning at sea is a bad thing. Seawater would enter the cabin, and when the cabin was near full the boat would be dragged to the bottom of the sea, pulled by her 10,000 pound lead ballast in the keel. We kept the main hatchway closed so that if we did overturn, that same lead ballast would serve to right the ship before too much water got into the cabin. We might survive a knockdown, and hopefully put Vonnie T back on a sound track. It would require an extreme amount of good seamanship to survive.

In the cockpit, we were both equipped with Henry Lloyd rain gear with built in safety harnesses. We clipped on so that we could not be washed overboard.

The auto pilot was useless in that heavy seaway, which forced us to take turns hand steering. I well remember how exhausting it became due to the intensity of the situation as the seas built higher and higher throughout the night. We had to keep going straight down those waves and not let them turn us. Paying attention was tantamount. We were soon so tired, that the shifts at the wheel were only 20 minutes long because that's all either of us could endure. The person who was not at the wheel lay down on the cockpit floor while still wearing the safety harness and went directly to sleep. And so it went through the night. (Yes, the bad stuff always happens at night) As the sun began to rise the next day, the winds and seas finally began to lie down. By early morning, conditions returned to 15 knot winds, and 5 foot seas.

We were now able to begin normal longer shifts at the wheel. I had the early morning watch shift, and LaVonne was fast asleep in the bunk. We each needed to catch up on our sleep. I thought about the long and arduous night at sea. My good wife stood by my side through it all. Without her, I wouldn't be writing this. I don't know where she got the strength and will to do what she did. She seemed to always rise to the occasion in difficult situations. She always helped me think my way out of trouble. She always treated me with kindness, respect and love. No man has ever had a better wife.

(Phew) We sailed into Cartegna harbor knowing that we had once again faced the grim reaper, and won.

We stern tied to a pier at a Marina near the center of town. We put out a bow anchor, and two lines to the pier. We needed a plank from our stern to the pier to get on and off of Vonnie T. Every day women greeted us with baskets carried on their heads full of fresh fruits for sale. The bus ride to the center of the city took only about 10 minutes. The historic walled in city of old Cartegna had survived many years of attacks from the French and English Conquistadors.

Beer, very good Anglia was only about $6 per case.

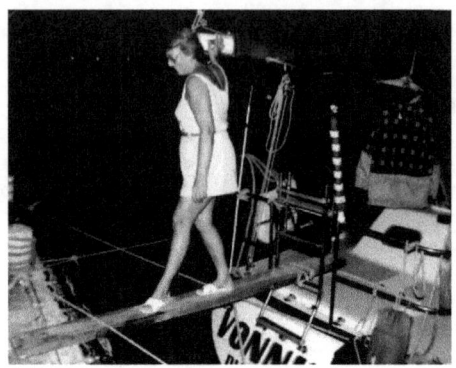

What a special place Cartegna was. They had a University there before the Pilgrims landed at Plymouth Rock. They had many beautiful people, many beautiful restaurants, and many beautiful historic sites. The streets in the walled in city were paved with bricks that were used for ballast in the old sailing ships. Those ships came to the new world with bricks in the bilges for stability, and went back with the riches of the new world. Some of the British plunderers brought bricks from Glasgow Scotland. Those bricks still pave the streets. The early plunderers arrived in 1533. They took a lot of gold that they had pilfered from the indigenous people, and a lot of gold that they had taken from the graves of indigenous people long since dead. The families of those people honored the dead by burying nicely crafted gold pieces with them.

Cartegna had a Church that dispensed inquisition punishment. At the time that inquisitions were going on in Spain, so too it happened in Cartegna. The Church was now a historical museum. Some of the torture devices were duplicated and on display. The Church had a slot along one wall where the faithful could accuse whomever they pleased of being infidels. The priests would open the accusation paper, and proceed to take into custody whoever was accused. The way that they got the truth out of the accused was by torture. I won't describe the torture here, but it was very gruesome. It was all in the name of Religion. It was not much different than what the Muslims did in Persia. It was not much different from the terrible violence that they are doing in 2015 against a sect of Muslims that they are at odds with. Yes, one sect of Muslims believes that Mohammed's followers should elect the heir to the prophet Mohammad. The other sect believed that the heir should be the next in line in Mohammad's family. These two sects are called Shiites, and Sunnis.

I found a hardware store that had bronze check valves. They did not have bronze hose adaptors, so I bought aluminum ones knowing that electrolytic corrosion was a possibility between the two dissimilar metals in salt water. But, I didn't know how long it would take, and I planned to replace them at a later date. I installed the check valve, and felt better knowing that a following sea could no longer get into the engine.

After fighting with the Auto Pilot that came with the boat, I scrapped it and bought an Autohelm 4000. This little guy operated from a big pulley and belt behind the wheel connected to a servo motor that was directed by a gyro compass. It was inexpensive, and when I thought about how I would repair it at sea, I just bought a second one for backup parts. The Autohelm operated on less that $\frac{1}{2}$ ampere. It wouldn't wear down the battery significantly. That Auto Pilot would make the long sea passages that were to come, a lot easier than hand steering.

I checked over our refrigeration, and made sure it had a full charge of Freon. The refrigeration system consisted of a 'Freon' compressor, and a cold plate in the refrigerator compartment, and another in the freezer compartment. The compressor ran off of 110V, and was cooled by seawater. The compressor compressed the Freon until it became a liquid, and then let it expand into a gas after passing through the expansion valve. As the Freon gas expanded, it took in heat from the surrounding space. The expansion took place in copper tubing that was placed inside the cold plate. The cold plates were about 3" x 12" x 16". They were stainless steel, and completely sealed. On the inside of the plates was the copper tubing, and 'eutectic' solution. The eutectic solution filled the plate cavity. In the freezer compartment, the eutectic solution was set to liquify at about 20 degrees. Different liquids freeze at different temperatures. Water freezes at 32 degrees F, antifreeze (ethylene glycol) freezes at about 8 degrees F. Water and ethylene glycol were mixed in the right proportion so that the resulting mixture, called a eutectic solution, had a freezing point of 20 degrees F. The compressors job was to cool the eutectic solution down below its freezing point so that it became a solid. Now this solid mass had a temperature of 20 degrees F. The compressor was shut down. Over a period of 8 to 10 hours, the eutectic solution would gradually thaw out. The heat for the thawing process was taken from the food contents of the freezer. A similar process happened in the refrigeration compartment, but that eutectic solution was set to freeze at about 35 degrees F. How cool is that!

In Cartegna, we discovered a place to buy plastic bags tat were about1 ½" in diameter, and about 12 " long. I had fastened 5 or 6 small brass hooks inside the freezer compartment, and immediately above the eutectic plate. We filled the little bags with water, tied them off, and hung them on the little hooks. After a while we had ice. We were now one of the very few cruising boats who had ice.

We found a fruit stand that had perhaps 12 different kinds of tropical fruits. They made smoothies out of the fruit. They gave us the option of mixing in milk, and sugar if we pleased. For a few coins we could buy a 12 oz cup of that delicious stuff. Sometimes we got two cups each. We tried all of the different fruits.

While we were there, the drug lord Pablo Escobar was imprisoned rather lavishly in his own estate. He had earlier tried to make a deal with the government. He would pay off the Columbian national debt of $10,000,000,000 in exchange for them forgetting the whole thing. That didn't work, so he had to stay at home..........

We spent two months in Columbia, traveled all over by bus and airplane, and thoroughly enjoyed the country.

San Blas Islands, Panama

The Isthmus of Panama lies east and west. The canal actually runs from north to south. About 10 miles north of mainland Panama lay the San Blas. They are a part of Panama, but are inhabited by Kuna Indians.

Let me digress. The inhabitants of the Americas were not always of European decent. The Europeans first came here in the min 1400's. Prior to that the people here were 100% Indians. Those peoples first came here by migrating across the frozen bearing straights from what is now Russia to what is now known as Alaska. They were following their main source of food, the Wooly Mammoth. Their migration started about 18,000 years ago. Those Indians, as we call them, now populate much of North, Central, and South America.

The Kuna Indians live today pretty much as they lived 1000 years ago. Their buildings are thatched huts built from branches. Their boats are hollowed out logs. Their food is fish, and whatever they can grow. Some food is brought from the mainland. They make the 10-mile crossing to the mainland in the hollowed out logs. We saw it happen.

The thatched huts always come in two, and protect families that may number into the twenties. One hut is used for sleeping (they use hammocks, and string them up at nightfall). The other is used for cooking and eating.

These are marvelous and simple people.

One of the things that the women make is 'Molas'. A Mola is a stack of six or seven brightly and differently colored materials. Slits are made in the first layer, and the hems are hand sewn. Each underling layer is slit to a design, and hand sewed in some meaningful pattern by hand. The Molas may be 12 to 24 inches square. They sell the Molas wherever they can. We had dugouts come by the VonnieT regularly. We bought them for $4 each, and had about ten of them when we needed no more. We were going to refuse to buy more, but thought "we are in their waters, we owe it to them to buy them" So we ended up with quite a collection. Later, we would find those Molas selling for $20 or more in U S gift shops.

One lady came by the boat with Molas, and a little girl and a little boy. The boy was nicely dressed and clean. The girl looked pretty much uncared for. LaVonne made a fuss over the little girl to show that she too had value. The lady held up the little girl and said "Quiere Usted?" That meant, "Do you want her?" Wow, we were taken back. How could this be happening? We knew that we couldn't take her for legal reasons. We were very sad. In retrospect, this woman was not offering her daughter to get rid of her. She was offering her daughter to give her a better life.

OK, let us go through the Panama Canal.

The Atlantic and Pacific oceans are at about the same levels. The landmass between them is some 90 feet high. The canal is a series of three locks up and three locks down. At the top is Lake Gatun. The daily rains feed Lake Gatun. Lake Gatun, in turn feeds the three sets of locks on each side. An up bound ship waits until the first lock is at sea level. The low level gates open and the ship enters. The gates close, and water from Lake Gatun is piped in from above. The ship rises to the next level, the high level gates open and the ship transits to the next lock. The same thing happens on the way down. Every time a lock is opened to let the water inside to escape out to the downstream, it must be replaced by water from Lake Gatun.
A 'Panamax' ship is 1000 feet long, 106 feet wide, and 41 feet in draft. That is the largest ship that can transit the canal.

We waited outside the first lock near Colon Panama. We needed permits and had to pay fees. Yachts such as ours are only allowed to pass about once per week, when that happens a bunch of us goes through at the same time rafted together.

Vonnie T leading a freighter

We chose to split our passage, and stayed in the middle of the canal at Lago Mira Flores for about two months. We stayed because we wanted to leave the boat for extended travel around Panama, and to make a trip home to MN.

We moored in Mira Flores in a kind of 'Yacht Club'. There were a bunch of boats there all touching a bit of dock. We tied to the dock, and then ran lines out across the little inlet to further secure the boat. We saw several alligators there, basking in the sunlight. While there, we met the last U S Panama pilot. The canal was in the process of being given to Panama, and they wanted only Panamanians working there. This Pilot was an interesting guy. He would come to our boat to just sit in the cockpit, and watch the big ships transit Mira Flores. He was 'in love' with the canal. I asked him one day when he planned to retire. He replied that he could retire now at nearly full pay. He said "but then I would be an ex Panama Canal Pilot" It was a high status thing to him. He did take LaVonne on a passage through the canal on a freighter. She later wrote about it in her book 'Andrea'.

When it was time to leave, we did the standard checking, and stowing. I undid the lines we used to tie up. We motored slowly out avoiding the other boat's tie up lines. Then it happened. A line got fouled in our prop. I quickly put the transmission in neutral. I knew what I had to do, and I remembered those alligators on the bank. I put on mask and fins and went over the side. The visibility here was poor. If an alligator came, I would feel it before I saw it. The line came free of the prop easily, and I was back on board in less than two minutes. I counted my legs, and yes, they were both there.

We passed through the remaining locks, and entered the Pacific Ocean. We sailed about 10 miles off shore to Islas Las Pearlas. Here we anchored and began the task of cleaning the bottom. Marine growth called barnacles grew all over the bottom. We had been in a cycle of hauling out about once a year to clean the bottom and repaint it with two coats of good antifouling paint. In between haul outs, we hand cleaned with scrubbies, and putty knives.

I had a $\frac{1}{2}$ hp compressor, 50 feet of hose, and a secondary regulator. The barnacles were pretty thick. Two months in the bad water of Mira Flores helped the little fellas grow. Soon we were 'ship shape' again, and began our voyage to Salinas Ecuador.

Ecuador and more South America

We had read about the Yacht club at Salinas, and knew that we could haul out and do a good job on the bottom. We also knew that we could travel to Quito, the capital of Ecuador and get a permit to visit the Galapagos. At the Salinas Yacht Club, we were assigned a mooring buoy. There were no slips. We were given guest privileges that meant we could use their restaurant. There were two classes of people there, The 'Socios' or members, and the Marinaros and other help. Ecuador at that time had about 60% of the population working for 'minimum wage' which was $35 per month. The cost of living was low there, but really!

Vonnie T was hauled out, and I hired one of the Marinaros to help with the sanding and painting. We put a ladder up the side, and continued to live on board. A few days later, we were back in the water. We watched as the children of the Socios came down to play with their toys. They had jet skis and a variety of small sailboats. The Marinaros job was to get the toy out of storage, put it in the water, and help the kid get underway. I couldn't help but think about the huge gap between the rich and the poor. The $35 per month Marinero was putting a $10,000 jet in the water for a pampered kids play. We left the boat at the mooring, and flew to Quito to get the Galapagos permit. In Quito, we visited the Capital building. In one chamber of the building, the walls were lined with paintings of the current Senators. I noticed that they were all of Spanish heritage. Equador is about 25% people with pure Spanish blood, 50% Indigenous people, and 25% Mestizos, or mixed blood. We wondered how it could be that the rulers of the Country could be all of Spanish decent. There was a guard on the premises who was Indigenous. I asked him why the Senators were all of pure Spanish blood when most of the population had Indian Blood. His reply was that he didn't know. There is some form of Democracy there, and apparently the people with Indigenous blood didn't participate too much in the voting.

We had been getting to know some of the Marinaros pretty well. One particular man seemed very trustworthy to us. We wanted to take a 2 month trip south into parts of South America that we had not seen. We were concerned about leaving the boat alone unattended for that long. I asked this Marinero if he would like to sleep in the cockpit every night to make sure the boat was safe. I offered to pay him $35 for each month, and a bonus of $35 if the boat was in good shape upon our return. He was delighted to get a chance for the extra money. On the day we left, I brought him out to Vonnie T, and asked him what we could set up for him in the cockpit. He insisted that we lock up the entry to the cabin. We left him some fruit, a blanket and a pillow. I felt real good about this. I knew that the other Marinaros would also keep watch to be sure that their buddy would qualify for the bonus.

Off we went with our back packs. Our trip would take us to Peru, Chile, Uruguay, Paraguay, Argentina, and Bolivia. We found that flying within a country was much less costly that flying from country to country. In Peru, we flew to the border, and then took a taxi across the border to Arica. Arica gets **no** rain. Arica is a mining town. The water is hauled in by truck and train from the south. The town looks pretty green, with lots of trees. We left Arica on a bus. Immediately out of town, we had a vista that had no animals, no plants, and no insects.

Arica, Chile   desert

We had a splendid trip. Who can say what the highlight was.

Valpariso, Chile

Santiago, Chile

Puerto Mont, Chile

Montevideo, Uruguay

Buenos Aires, Argentina

La Paz, Bolivia

And many cities and towns in between.

We entered La Paz Bolivia on a bus, coming from Argentina. As we came into the city we passed soccer fields, where the youth were busy at their sport. We were at 13,000 feet altitude, and knew we would have trouble walking at even a slow pace, but here these guys were playing soccer.

We found a hotel near the center of town, and walked to the center for dinner and a look-see. It started to get dark, and we headed for the hotel a mile or two away. Then the lights went out in the whole area. We were walking back in the dark, not knowing exactly how to get to our hotel. I thought that if ever we are going to get mugged, this would be the time. As we walked we stopped a couple of times to ask directions. Those people were very helpful. When we got to our hotel, the porter was waiting for us at the door with candles.

Lake Titicaca was out next stop. It lies at 13,000 feet above sea level, and is the highest fresh water lake in the world. It is big too.

The reeds stacked neatly by the shore are the same type that Thor Hyerdahl took to make Kon Tiki many years before.

A Beautiful Bolivian Boy

Now back to Salinas. We found the boat perfectly in tact, if fact it was cleaner than when we left it. We did some final provisioning at the market in Guayaquil. Again every thing was cheap and plentiful. We bought bananas and oranges, and some very nice cuts of meat. Next thing was to take on the last fuel and water that we would have access to for a long time. Four five-gallon jerry cans were lashed on to each side. One set was for extra water, and the other was for extra diesel fuel. Water was in short supply there, so we went to the local bottled water plant and bought about 50 gallons at $.10 per gallon. We topped off our tanks, and filled the jerry cans. We also topped off with diesel fuel at $.40 per gallon. Now we were ready.

The last thing was to help out the Marinaros. When they had to take a day off for being sick, they were not paid. We gave the secretary of the yard $100, and told her to start a fund for use by any Marinero that needed sick pay. Wow, was that a hit! She quickly got on the walkie talkie, and told the guys what we had done. Within minutes, every one of those fine people came to the office to thank us.

We stood tall that day.

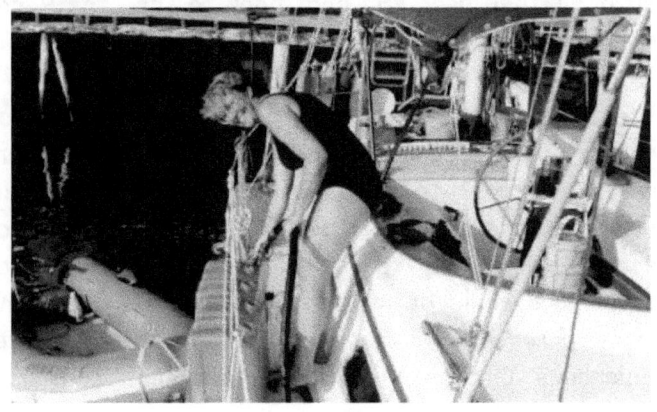

The South Pacific

We gathered our Zarpe from the port captain's office and made way for the Galapagos. It would be a 600 mile trip, and should take us about 4 days. Two days out, the wind died and we were becalmed. It was very hot. We took down the sails, and decided to take a swim. There was no land in sight. I swam leisurely about, and was soon 50 yards away from Vonnie T. I thought, 'what if the wind came up?'. If the wind came up it would push our boat to leeward even without sails. Would I be able to swim fast enough to catch up, or would I become a person lost at sea? I swam immediately back, got in the cockpit, and waited for the wind to blow. Sure enough, in about an hour the wind began to build. We sailed nicely through the night. The next day my reel began to spin. I had a big one! As I reeled in some line, I saw that it was a pretty big shark. Why would he attack an itty bitty Rapala CD 14? I knew that I wouldn't be able to buy another lure like that any time soon. The battle was on. After 55 minutes, I was able to bring him up to the stern, and belay the leader around the stern cleat. I then used the boat hook to free the Rapalla. The shark swam slowly away, but I had my bait back.

We entered the Galapagos one fine afternoon, and began our anchoring procedure. I soon learned that the transmission was not working properly. It would engage at slow speed, but as I increased the throttle, it would slip. We did the anchoring at slow speed. Now what to do? We had the entire Pacific Ocean to cross. It would be nice to have a reliable transmission for the many ports that we would have to enter. At anchor in the Galapagos, I tried to line up a replacement transmission to be sent here, and installed by some mechanic that I was yet to meet. I was not able to get a transmission. Now the dilemma: Would we turn back or go forward? After a lot of thought, we decided to go without a fully functioning transmission.

The natural beauties of the Galapagos are a thing to behold. We saw The mating dance of blue footed Boobies (yes their feet look like they have been dipped in blue paint), Galapagos turtles, Marine Iguanas, sea lions, penguins, the Frigate bird mating ritual, and on and on.

The Frigate birds were particularly noteworthy. The male builds a nest in the tops of some brush. He then inflates a bright red sac in front of his throat. There may be several of these nests, all with competing males. Meanwhile, the females circle overhead trying to decide which male would make the best mate. Once her decision is made, she dives on the male's nest, and appears to be hugging him as their wings intertwine.

One night in the Galapagos anchorage, a small boat came alongside. One of the men climbed into the cockpit. I immediately scrambled up, and was irate to think someone would come aboard without permission. The three guys were dressed up in Ecuadorian army uniforms.

I got right in the intruders face, and told him in my best loud Spanish to "GET OFF OF THIS BOAT NOW". It worked and he retreated to the small boat. Now we could talk. They asked for and anchoring 'fee' for using the harbor. I explained to them that we had a permit from the Ministerio de Defensa, and that there was a copy of it in the port captain's office. I further told them that we would pay them no fee. Another attempt at extortion!

The day before were to leave we got a very nice rain. It almost never rains in the Galapagos.

With that welcome rain we could top off our water tanks I had installed a Y valve under the starboard scupper. When directed the normal way, deck water would pass back into the sea. When directed the other, the water would go into our holding tank. We scrubbed down the deck to rid it of salt water. I then tested the discharge into the Y valve by tasting it. When it was no longer salty, it could be diverted to the holding tank. We counted this as a small miracle. We weren't sure when the next opportunity to top off the tank would occur.

Crossing the Open Ocean

It would be 3050 nautical miles from the Galapagos to the Marquesas. When you are half way there, you are the farther away from land than at any other place on earth. Looking at the sky, we could imagine an arc that was 1/7th of the distance around the world. That would be our route. I would take us 21 days to do the crossing.

We had made only a few multi day passages until now. We would need to have around the clock watches to keep Vonnie T out of trouble. The Navy does their watches 6 hours on and 6 hours off round the clock. We tried that once and found it didn't work for us. We would both get tired at about the same time, and both want to be awake at the same time. So we worked out a simple system: If you could sleep you did so, and if you needed sleep, you woke the other person. LaVonne could fall asleep easily, and would usually go to the bunk shortly after dinner in the evening. We planned our dinners so they coincided with running the generator, and occurred in the twilight. Many nights, I was on watch before first light. That first light is something special to watch at sea. The horizon first gets a faint orange color, and then the sun pops up soon afterward. Full moonlit nights are also great. You sit in the cockpit hour after hour and watch the moon slowly pass overhead. Moonless nights are pretty good too. In the darkness of the open sea, and away from any land lighting, the stars seem to jump out of the sky, and you can see millions of them.

Vonnie T was set up pretty nicely to be independent for extended periods of time. The sails were all we needed for propulsion. We had an 8 KW Generator to make 110V power. We ran that

Generator twice per day for about 40 minutes each time. With that power, we charged the batteries, cooked in the Microwave oven, ran a hair dryer, heated water in our small electric hot water heater, and replenished the 'cold' in our freezer and refrigerator.

At sea, we wanted to go a bit south to avoid the doldrums that occur near the equator. The Galapagos Islands lie about 40 nautical miles below the equator. We wanted to be 6 or 7 degrees below the equator, which would put is in the path of the north east trade winds.

The Autopilot did all of the steering.

LaVonne had planned ahead to grow lettuce on our crossing. She had two plastic dishpans with perforated bottoms for drainage. She filled them with the best dirt we could get, and planted lettuce. As I recall it took only a few days before the lettuce was first big enough to eat. Every time she picked a row of lettuce, she would replant. Let me tell you, when you are days at sea fresh lettuce is pretty darn good. She also made bread, using the engine compartment to let the bread 'rise', and then baked it in our oven. Wow, love that woman.

We had gentle 10 to 15 knot winds coming from abaft of beam that gave us a nice broad reach. We made good speed, and conditions were very comfortable, about 90 degrees in the daytime, and 75 at night. The seas were running 4 or 5 feet in height. We caught fish, and did a lot of watching the water, and reading. I even practiced my celestial navigation. Life was good!

One day we heard a familiar voice on the VHF radio. It was Neil and Scarlett on the vessel Lebenho. We had spent many pleasurable days with them in the Caribbean, and now they were only about 30 mile north of us and headed in the same direction. What a coincidence. They had come through the Panama Canal too, and headed straight for Hiva Oa.

On the 21st day we sighted land. Our Sat Nav told us it was the outer Marquesas. I remember not having boat fever. We did not desperately need to get off of the boat and onto land, however we wanted to see Hiva Oa. We set the anchor in the harbor, and did our routine boat cleaning and set up. We had averaged 5.9 knots on the 3050 mile passage. I got into the engine compartment and checked the engine oil, and transmission fluid. What a surprise. There was almost no transmission fluid on the dip stick. In the past, I had always checked the transmission fluid with the engine running just like I did with automobiles with automatic transmissions. Then it hit me. THIS WAS NOT AN AUTOMATIC TRANSMISSION! When the engine was on, the transmission gears still rotated throwing oil, as low as it was, onto the dip stick. I had been servicing that transmission wrong for over 3 years. Well, I put a couple of quarts of transmission fluid in the bugger, and guess what? The transmission worked fine again. I thought to myself: 'You Dummy'.

We inflated the dinghy, launched it, put on the outboard motor, and went ashore. We were surprised to find how weak our legs had become. After almost no walking for 21 days, our leg muscles got weak. We took it easy at first, but by the next day we were back in shape and eagerly exploring this South Pacific Island.

Next harbor was in the Tuamotu Archipelago. It is an Atoll.
An atoll is a land mass caused by the eruption of an underwater volcano. As the volcano erupts, it brings bits of the earth upward, and they spread out in a pretty big ring. If the volcano continues to grow, a land mass is created. The Hawaiian Islands are examples of the tremendous power of the volcano. The Atoll is formed from the earthen stuff heaved upward, and into a circle. The earthen stuff may stick a few feet out of the water, or perhaps as much as twenty feet. The volcanic activity subsides, and what remains is this large ring of volcanic ash. Inside the atoll the waters are calm, outside the ocean continues to do its rolling action. Tides affect life inside the atoll. As the tide recedes from a high, the water inside the atoll finds a weak place in the volcanic ring, and erodes a channel from inside to outside. Likewise when the tide goes up, the water rushes in through that same opening. The currents formed by the tide can reach 6 or 7 knots. Boaters learn to time their passage through the channel so as to take advantage of the current.

We went to the atoll named Suwarrow, in the northern Cook Islands
Suwarrow is an outpost in the northern Cooks that is 400 miles north of the Central Cooks. Suwarrow was populated by eleven

people who pretty much lived off of the land and sea. They get a supply boat twice a year. Their principal duty is to occupy the land there to maintain its tie to the lower Cooks. They have a customs and immigration officer. We checked in with him after anchoring, and found that we were very welcome to anchor there. The officers name was Takaro, and he was 21 years old, and wore no uniform, only shorts, and bare feet. I quickly became friends with Takaro, and learned that he was an avid snorkeler and spear fisherman. Takaro had only a small boat with no engine to propel it, so his forays into the lagoon were pretty limited. With my dinghy, we could whiz across the water to very good diving spots. My first dive with him was in pristine water teeming with fish. I went down to my comfortable 35 feet, and I noticed that Takaro kept on going to perhaps 50 feet or more. We shot several nice fish, and Takaro noticed that a few Sharks had become interested in our activity. We decided to hunt elsewhere.

Takaro had a way of dealing with lone Sharks. One day when LaVonne was with us, we were all in the water hunting and sightseeing. Takaro had shot a fish, and tied it to a stringer he had around his waist. Now this is something that I would never do. If a Shark went after that dead fish, he would likely get a piece of me too. However, Takaro had coexisted with Sharks for some time and knew how to deal with them. Sure enough a small white tip Shark about 5 feet long came to check out Takaro's dead fish. The Shark got pretty close when Takaro pushed him away. The Shark returned and Takaro made a fist and hit the Shark in the nose as hard as he could. The Shark got the message and swam away. LaVonne, not wanting to have any part of a Shark fight, got into the dinghy.

Back at the anchorage outside the little village, we met all of the people who lived there. They were mostly one family, and lived a very Spartan life.

We brought gifts of canned goods and fishing supplies. They were happy to have contact with the outside world. Soon other boats began to arrive. I think there were about 6 cruising boats there. We learned that the last time a supply ship came to Suwarrow, a disease hit the family. Everyone got sick for a week or more. We were later to learn that Dengue fever was the sickness.

We spent a week or more there enjoying our new friends, and the other cruisers. When it was time to go, I readied Vonnie T, and LaVonne took the garbage to the dump. We were the first to leave one sunny morning, and the herding instinct forced the other boats to follow us. We were a bit faster that the other boats, and arrived in American Samoa after a four day sail.

American Samoa became 'American' during WWII. The U S wanted a submarine base from which to attack The Japanese fleet. The harbor at Pago Pago (pronounced Pango Pango) was protected and well suited. The U S government made some sort of deal, and the Island became a territory of the U S. Years after the war, the Navy vacated the harbor, and to leave the Islands people a source of income, they facilitated turning the harbor into a tuna fishing port complete with two large tuna canning plants.

We checked in, and took a mooring in the harbor. Before long, LaVonne was in bed complaining of aches and pains. The next day, the other boats came trickling in. Each of those boats had a sick person aboard. It turned out that the sick person was the one that took the garbage to the dump. They had all been bitten by dengue fever carrying mosquitoes. Each of those other boats was quarantined, and told not to come ashore until permission was granted. Well, we were not quarantined, so I became the person that shuttled between each boat and the shore bringing whatever the crew of the boat wanted. I brought pizzas, fresh milk, newspapers and etc.

Dengue fever is a pretty bad sickness. It is similar to Malaria. It brings high fever, aches and pains, and a very weak body. It took LaVonne seven or eight days to get strong enough to want to go ashore. Even then she was anxious to get back to the Vonnie T

193

and lie down. One of the boaters got so bad that his blood white cell count was alarmingly low. Our doctor friend, Gene Eisenberg who was also a cruiser, recommended that he be flown to Hawaii where he could get a blood transfusion. In the end everyone got healthy, and proceeded with their adventures. Gene had visited the Hospital there in American Samoa, and reported that it was very humble. In fact it did not have screens on the windows to keep out the mosquitoes.

We next visited Western Samoa, which is an independent nation. We checked in at Apia, and had a delightful stay. We sailed down the coast looking for a safe and unique anchorage. We found one outside of what we later found out was the Piula Theological College. We tried to anchor there using our normal 'back the anchor down' procedure, but it wouldn't hold. I had to dive in about 12 feet of water, and wrap the anchor chain around a big coral head. It turned out that we were the first cruising boat to anchor there. The next morning, we heard marvelous music coming from the Church on the shore. We went in with the dinghy, dressed in our finest Lava Lava's. We wanted to go to the service, and listen to the music, and meet some of the people. We were greeted on the shore by a sentry from the church. He told us that this was private property, and we weren't allowed. We explained that we just wanted to go to the service, and suddenly we were welcome. It was a large church full of perhaps 300 people. The music was great. After the service, the preacher, who turned out to be the Dean of the college, and his wife came to talk to us. We had rather stuck out being the only white people in the group. The Dean had married a woman from Iowa. She was delighted to have Americans visit. We were invited to participate in their Sunday feast. The meal is prepared the evening before in an 'Umu'. The Umu is a shallow pit that is partially filled with large stones. On top of the stones a fire is built to heat them. The fire goes out, and banana leaves are placed on the hot rocks. The food is then placed on the leaves, and more leaves are put over the food. Lastly, the Umu is covered with dirt. The cooking process takes place over several hours, The Umu is not opened up until the next day.

We were led to a ceremonial 'Fale' that was used for chieftain meetings, feasts, and important events. Fales was the name they gave to their buildings. Fales had roofs supported by poles, a thatched roof, and open sides. We were part of a group that consisted of all of the professors from the college. We each sat with our backs against a support pole. The food was placed in front of us. There was a lot of food. We each had a banana leaf to serve as a plate. The students of the College served us. I was served with enough food for five people. I didn't understand, but did not want to give insult by refusing some of the food. There was fish, poultry, and a nice variety of vegetables. All of it was good. As it turned out, the custom was to offer the faculty and their guests the first shares of food, which was way too much. Then after the group ate, they left, and the students ate the rest. What an experience!

Umu

Feast

The next day after meeting many of the people, they found out that I was pretty good with electrical repairs, and asked me to take a look at a few of their problems. I went back to the boat and returned with tools and a bunch of electrical parts. First they showed me an electric fan that was dead. I took it apart, and noticed a broken wire, which I easily spliced together. Next, a man had a broken TV. I told him that I knew nothing about TV's, but he insisted that I take a look. I took the back off of the TV, and noticed a fuse. I inspected the fuse, and found it to be burned out. I searched through my little bag of parts and voila, found the exact fuse that was needed. We plugged in the TV, and it worked. I was regarded as a hero. Another lady had an electric iron, and so on. I felt really good by being helpful to these kind and simple people. They found out that we had good cameras, and asked if we would take class photos. That would be easy. We worked out a time when everyone would assemble. We returned to find them all dressed in very clean white garments. Each class posed on the front steps of the church. We took photos of each class, each class with their wives, all the kids together. We took dozens of shots, and told them we would get them developed in New Zealand, and mail back the prints and negatives.

We had a marvelously rich experience in Western Samoa!

SCUBA

Self Contained Underwater Breathing Apparatus

I had SCUBA gear on board, but rarely used it since I had no compressor to replenish emptied tanks. The tank straps onto ones back. It holds air at about 3000psi pressure. It had a primary regulator that cuts the pressure down to about 100 psi before delivering it to a hose that terminates in a secondary regulator, and a mouthpiece. That secondary regulator is an 'on demand' device that only lets air into the mouthpiece when the diver draws on it. The diver also wears a BC or buoyancy compensator. The BC is connected to the tank with a hose that has a valve that the diver can inflate the BC or let the air out of it. The diver also wears a weight belt to make it easier to go down. The diver starts out with his mask, fins, and SCUBA gear in place. He puts some air into the BC, and puts the mouthpiece into his mouth. He then jumps overboard, adjusts everything, and lets the air out of his BC. He then slowly descends, and starts his exploration.

I used SCUBA when we were in the Kingdom of Tonga. There was a dive shop there that would fill my tank.

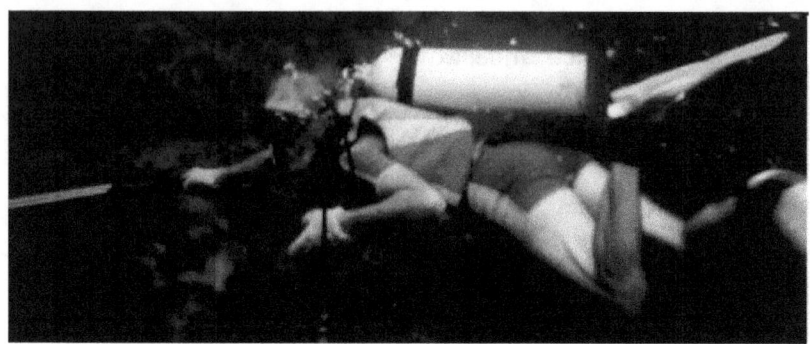

Now this is a story of water clarity. I had on my SCUBA gear, but was breathing through my snorkel, and cruising along the surface looking for fish. I spied coral formations on the bottom, and several nice looking fish. I switched to the SCUBA mouthpiece, and let the air out of my BC. I had a depth meter so I could see how deep I was diving. Soon the depth meter read 100 feet! I was not on the bottom yet. The visibility in that water must have been 120 feet or more. I always preferred to dive in water 30 to 60 feet in depth. At 100 feet you use a lot more air, and can only stay down 15 or 20 minutes. Also, if you stay at 100 feet for any length of time, you must come up slowly to avoid getting the 'bends'. I retreated to the surface slowly, and would look for more shallow waters to hunt in.

When you are ascending from any depth it is important to continue breathing normal. If you are at a depth of 33 feet, the hydraulic pressure of the water is about 15 psi. Now, ones lungs are not a pressure container. The regulator at 33 feet puts 15 psi air in the lungs to match the hydraulic water pressure at that depth. If you are at 66 feet, the secondary regulator feeds your lungs 30 psi air. The pressure in your lungs is equalized with the water pressure at depth.

Bad things happen if you ascend from the depths, and do not exhale. The pressure in your lungs remains high since it was introduced there at depth to match the hydraulic pressure of the external water. The rising pressure in the lungs causes them to expand. That expansion can result in rupture if not relieved.

Tonga

In Tonga there is a place called 'Mariners Cave'. It is well known to boaters, and is written up in the cruising guides. To enter the cave, one must dive to about 8 feet and go in through the opening there. You then glide up to the surface, and see the wonders of this cave, which is lit through a natural skylight. I had no problem snorkel diving to enter the cave, but one of our fellow cruisers needed to use SCUBA to enter the cave. On one passage, I asked him to let me have a breath of air from his tank at a depth of 8 feet. I got the air, and started to the surface but I FORGOT TO BREATH OUT AS I ASCENDED. The pressure in my lungs slowly rose, and I felt the discomfort. The light bulb went on in my brain, and I quickly exhaled to relieve the pressure. Had I done that at perhaps 30 feet, and not exhaled, I would not be here to write this.

In Tonga, we anchored near a small Island that had perhaps 200 inhabitants. There were two churches there, and I got to know one of the preachers (yes, we went to church for the music). One day I noticed a few men digging in the cemetery that was adjacent to the church. I asked the preacher if someone died. He said "no, it is just the Tongan way". I asked again and got the same response. Then finally he told me that people from outside Tonga couldn't understand. He pointed out one of the men digging, and said the man's brother had died a year earlier. He said that the living man had been experiencing a mild pain in his side and he thought it was a signal from the dead brother that something was wrong in the grave. The men dug up the corpse, and wrapped it in fresh Tapa Cloth. They then smoothed out the bottom of the grave to be sure that no stones or sticks would bother the dead brother. Wow, how can rational people believe such a thing? Well, maybe some of the stuff we believe in is also a bit irrational.

There was to be a 'Fete' on a small island near by, so we went over there and tried to anchor like we normally did. However, the place near the Fete was not a good anchorage. It was all down slope into deep water. Our anchor would not stand the test of being backed down at 1/3rd throttle. Since we were going to be there only a couple of hours, I decided that the anchoring would be OK with only minimal backing. Well, the splendid Fete went on and on. Soon it was near dark, and we decided to stay put for the night. While we were sleeping, the wind shifted and grew in intensity. I awoke knowing something was wrong from the boat motion, and the sound of waves on the hull. We were adrift with the anchor and chain hanging straight down. We were being driven stern first towards a jagged and rocky cliff. I gave a loud yell to LaVonne, and scrambled to start the engine. Soon we were back in control again. We had missed disaster by perhaps 15 minutes. This was the first time in our cruising years that we had dragged anchor. We went to a safe anchorage.

Another day, we were headed across an area that had a few reefs, but our chart showed how to make the passage. We set the course at 240 degrees magnetic. Lavonne was at the wheel, and I took on some minor boat repairs. Suddenly, I noticed that we were headed toward one of the reefs. Not good. I pointed out the right way to LaVonne, and looked at the compass. It had stuck on 240 degrees. It was carded for the northern hemisphere where the earth's lines of magnetic flux are different. The card that lies in a bath of compass fluid and had been weighted to balance in the north. Now, in the south, those weights had caused the card to tilt, and it became hung up on something inside. Not to worry, we had hand compasses, and I lashed one to the binnacle. In between these times, we had many very pleasant experiences with the kindly Tongan people. Now it was time to head for New Zealand, some 1,000 miles to the south. We bought fuel, and topped off the water tank.

Leaving Tonga

The morning that we headed out we were given a special treat. I had my Rapalla bait out and a big, maybe 200 pound, Marlin took it. He leapt from the water, and did a tail dance before falling back into the water. Again he jumped out of the water and did his wriggly tail dance until the line broke. While I didn't like to lose fishing gear, the thrill of that dance was a wonderful Tongan goodbye.

About the second night out, we needed to motor sail to get upwind of an island. The engine would not start. I first tried to bleed the fuel injection lines, but that didn't work. No fuel would come out even with the lines loosened as I cranked the engine. I had a think on that, and would see if there is something to be done about it tomorrow. The next day I took the fuel injection pump apart, and found that the main shaft had been twisted in two. For sure, That Tongan fuel had a lot of water in it. Normally, our onboard fuel water separator would have removed the water, but not in that quantity. The water had entered the injector pump, and caused it to seize up. When we tried to start the engine, the shaft failed in torsion. Not to worry, we are a sailing vessel.

32°S   Latitude   174°E   Longitude,   16   Nov   1991

Dead boat! We had already lost our main compass, our engine and now this.

We were sailing in light air as close to the wind as we could. The wind was pretty much south, and that was the direction of New Zealand. In sailing close, I used the Autopilot and tried to stay about 50 degrees off of the wind. More than 50 degrees and we were losing progress toward NZ. Less than 50 degrees and the sail would luff up. If we went through the eye of the wind the sails would backdraft, forcing us to do a 360-degree turn. So if the wind shifted slightly towards our bow, I quickly turned off the Autopilot, and steered hard to the leeward. This usually worked if I was quick enough. As I steered hard leeward, the rudder would come up against stops that would prevent it from hitting the hull. I knew that, but never mind, when it came on the stops, I continued to apply a lot of needless pressure on the wheel.

The last time I did that, something snapped, and the wheel turned freely. I had broken the heavy #50 Stainless Steel roller chain that connected the wheel to the cables that turned the rudder! What a dummy!! Now we are hundreds of miles from NZ with no main compass, no engine, and no steering. We are a dead boat. Depression is a mild word for what I felt. The boat began wandering aimlessly, out of control, and the sails flapped all over the place. I gathered a little sense and took the sails down. I sat in the cockpit and looked at LaVonne, with my face nearly falling off. The day was nice with little wind or waves, so we sat there for some time having a 'think on it'.

The two of us liked attacking a problem together. We decided we better dig out the emergency tiller. It fastened directly onto the rudderpost. I had built it from a three-foot piece of oak two by two. It wasn't comfortable to use. LaVonne stood over the place where our stern bunk had been removed with one end of the tiller in hand. She stuck her head up through the overhead hatch as far as possible. She couldn't see very well, but we were barely moving so it wasn't important for now. This was a big deal, now we had a way, although clumsy, to steer the boat. Now we would see what could be done with the broken steering chain. I had repaired roller chain like this before with a new master link, but alas, that was one spare part we didn't have. After more hashing it out between us, I decided to make a master link by destroying two of the regular links to scavenge parts. I chiseled them apart, and fastened a keeper from a washer in our stores. With the chain now whole, we adjusted the length of the cable that was attached to it. The cable ran from the chain around the steering quadrant to the rudderpost. It sounds rather straightforward but each step had to be invented. The whole process took about two hours.

Hurray, we are once again a functional sailboat.

Our generator continued to work even though it was drawing fuel from the same tank that had the water in it. I guess the engine got the bulk of the nasty water. The generator allowed us to charge the batteries, run the refrigeration, cook in the microwave and etc.

Our cruising ham friends had been listening to our progress each morning on the 20 meter net. One boat that we had become close friends with offered to put us along side the customs pier when we got to New Zealand. A few days later, we entered the bay outside of Opua NZ. We were greeted by Bob and Janet aboard S/Y Jubilation. When we neared the channel entry point, both boats dropped sail. Jubilation came along side, and secured Vonnie T with lines and fenders. They then parked us at the NZ customs dock.

After clearing customs, we were allowed to stay there until we got the engine fixed.

New Zealand is a land with 3,000,000 people, 10,000,000 sheep, and 1.000.000 boats. There are a lot of marine services available. I took the injector pump and the compass into a marine shop. They had them both fixed within days at a reasonable cost, and we were back in business.

We anchored off in the harbor in front of Opua, and were reunited with many cruising friends who were also there to escape typhoon season. We bought an old Isusu car, and toured NZ. I can only say that New Zealand is a top five destination. The people are very friendly. The country side is wonderful with everything from farms to glaciers.

We took a two month tour in our little car. We visited the towns of Wharangarae, Aukland, Wellington, Christchurch, Dunedin, and a hundred towns in between. We ate lamb, Kiwi fruit, and fresh vegetables and stayed mostly in what they called 'Campgrounds'. In a Campground, one could choose to stay in a tent, rent a small cabin, or rent a little apartment. We met many fine people, and even had the pleasure of meeting a guy that had been in the National Sheep Shearing contest the day before. How many sheep can you shear in an eight hour period?

It was time for our annual trip back to Minnesota to visit family and friends. We found that the cost was the same to fly round trip to Minnesota, as it was to fly around the world with a stop in Minnesota. Air France did it for $1,300 each. You could take up to a year to do it on your own schedule, and get a side trip out of Paris to anywhere Air France flies in Europe. We thought "why not?" Next thing you know, we put Vonnie T up on the hard in Opua, left our car with the friendly mechanic, and loaded our backpacks for a new adventure.

Around the World

I remember first driving that rental car in Minnesota. In New Zealand, we drove on the left side of the road just like they do in Great Britain, and Japan. Well, after many months of driving on the left, now I must switch to driving on the right. Early on, I drove down the left side, and was surprised to see a car ahead turn the corner, and get in what was his right hand land. He was headed right for me. I quickly moved over, and would not repeat that mistake again.

Next stop: Paris France.

In Paris we found an inexpensive hotel just two blocks off of Champs Elysees. We had been to Paris before, but there is so much to see that we welcomed another shot at it. We thought it would be fun to get a car and Travel Europe, the Baltic Countries, and the remnants of the Soviet Union. The Berlin wall had just come down two months previously, and those old USSR Countries were just now breathing the fresh air that was denied them by the Russians. We made a deal with the Renault dealer. We bought a small new car with the proviso that the dealer would buy it back in two months time. It was like a rental, but because of some tax laws there sale and buy back was good for Renault. The cost of the car would be about $25 per day.

We would drive through a part of; France, Belgium, the Netherlands, Germany, Denmark, Sweden, Finland, Estonia, Latvia, Lithuania, Poland, Hungary, Czechoslovakia, Austria, Italy and Monaco. All are beautiful countries with unique cultures. I will write about some of the unusual things that happened along the way.

In Denmark, we needed to get the car serviced for its first check up. While waiting at the dealership, I strolled around and looked at the floor models for sale. I noticed that the prices were very high. I talked to a salesman there and was told the prices were high because the government tax on each sale was 100%. The $20,000 car actually was priced at $40,000. Now, we complain about high taxes, but the taxes in many countries are very high. Out of every dollar earned (or Kroner) in Sweden the total tax rate was about 55%.

We stopped at a gas station in Sweden, and after fueling, asked how far it was to our next stop. Note; most of the people in Sweden speak English. It was afternoon, and we always tried to get situated before dark. The service station attendant replied that it was a long way away, about forty miles. We thought that would be easy. We went down the road for hours before getting to our destination. It turned out that one Swedish mile was equal to 10 kilometers. 400 kilometers is about 250 statute miles, as we know them.

In Jokkmokk Sweden in early June, we experienced the 'land of the midnight sun'. Midsummer is June 21, but even in early June, we could read a newspaper outside at midnight. We had to work at going to bed at a reasonable time, say 10 PM. Jokkmokk is also part of Lappland.

June 21st is the summer solstice. On that day, the sun is the farthest north of any time of the year. It lies directly overhead if you are anywhere on the Tropic of Cancer (also called the Arctic Circle) which is 23 $\frac{1}{2}$ degrees north latitude. On that day, the northern hemisphere experiences its longest day. Conversely, on that day if you are in the southern hemisphere you experience the shortest day of the year. After summer Solstice, the sun appears to travel a bit southward each day until it reaches the Tropic of Capricorn, 23-$\frac{1}{2}$ degrees south latitude. In reality, the sun doesn't travel at all. It is we on earth that does the traveling. We orbit the sun once each year. While orbiting the sun, we spin on a slightly tilted axis. Each rotation of the earth about its axis is one day. As the earth orbits around the sun, it gives a slightly different face to the sun each day. The 21st of December is called the Winter Solstice, and brings the shortest day of the year to the Northern Hemisphere.

Lappland is another thing. Lappland is a vast area that covers most of northern Norway, Sweden, Finland and Russia. Long ago it was its own political entity. One day the four nations decided to extend their borders northward along lines of longetitude to the Barents Sea. The four nations still respect the ownership rights of the Lapp people who are nomadic, and indigenous. Those people herd Reindeer as their principle source of food, and income.

We passed into northern Finland, and were met with a new language challenge. There was no English, Spanish, Portuguese, nor French spoken. When we found a hotel, all of the exchange was done in crude sign language. When we were in restaurants, I would walk around the room until I saw something that looked good, and motioned to the waiter to get me a plate of that. We were always proud of the fact that Minnesota had 10,000 lakes. Well Finland is about the same size as Minnesota, and it has 50,000 lakes.

Here are pics of a Finnish Inn with our little Renault outside, and Reindeer.

We spent a few nice days in Helsinki before boarding a ferry for an overnight sail to Estonia. We met a young lady on the ferry who was a professional tennis player, and who had attended a University in the south of Florida. That was a great meeting. She explained to us in English that Estonia had only achieved independence from The Soviet Union two months previously. She showed us around the city of Tallinn in only the way a local could. She showed us the flat where she lived. She did not own it, but was assigned to live there by the government. The flat was in the family, and would go to the next generation when she passed. She told us that during the Russian occupation, many Russians came to Estonia, and were all given good jobs, and good apartments. We went to a grocery store there, and were amazed by the Soviet system. The store was not large by our standards. They controlled the number of people that were allowed to enter with shopping carts. They had about 30 shopping carts. If all of the carts were in use, one had to wait in line for someone to exit the store, and free up a cart. We got there at a slack time and went right in. Inside the shelves were about 1/3rd full. Compare that to a modern Wal-Mart!

 The currency there was converted from Russian Rubles to the new Estonian Kroon. They were very proud to have their own currency. It had been introduced only a few days earlier.

Latvia

We needed gasoline, so we found a station. In my best sign language, I asked for help pumping gas. The system was that you paid before being served. I had a supply of U S dollars, French Francs, and a few left over Estonian Kroon's. The lady didn't like any of my currency, and motioned me to go to a money changer. This is all in sign language. I approached the guy, and asked for a conversion from U S dollars to Russian Rubles (the currency of Latvia). We agreed, and I gave him a 20-dollar bill. He gave me a stack of Rubles about 1 $\frac{1}{2}$ inches high. OK, we got the gas, and I handed the cashier a small stack of rubles. She peeled off what she needed, and returned the rest to me.

Down the road we went. Soon the money changer pulled up alongside of our car, and motioned for us to pull over. We thought that might bring big trouble. I accelerated, and sped ahead of him. Again, he pulled even with us and motioned for us to pull over. Something inside of me told me that it was OK. I pulled over, and he explained to me in sign language that he had given me the wrong change. I handed him my stack of rubles. He counted, and peeled off about 75% of my stack, and returned the rest to me. We pieced it all together and decided that he had given me change for $200 instead of $20. We felt pretty good about having done the right thing.

Our daily practice was to start looking for a place to stay the night about 2 PM. If we found something quickly, it would leave plenty of time to explore the area. If we didn't find something quickly, we would have a few hours of daylight in which to do so.

We were traveling along a minor highway in Latvia, and were only about a mile from the Baltic Sea. We took a side road thinking that there might be a hotel or inn on the water. Sure enough, we sighted a pretty good sized building on the water. It turned out to be a hunting lodge that rented out rooms in the off season. We took a room, and began exploring. We, of course, were the only English speaking people there. We met some local people who told us, in sign language, that there was going to be a party in the basement of the lodge that night. We were invited. Someone showed us to the basement room, and what a surprise. It was a very large and well furnished room with a huge fireplace, tables set up for about 30 people to eat, and they even had a sauna. Music was playing on a recording device, and people were dancing. We joined right in. When we sat down, I was met by a series of women that wanted to dance with me. I guess it was because I was tall, and unusual. The meal was served by a committee of the local people. We tried to ask what the meat was, and eventually got the word 'Bambi'. We were eating venison. The group drank Vodka in straight shots. Many people took saunas. We had a delightful evening.

We passed through Lithuania without incident. The one thing that I do remember was staying in a humble cabin in a campground, and using a common bathroom. The cost of that cabin was $1.75.

Poland

We entered Poland and were pleasantly surprised to find it a pretty modern country. We wanted to buy some ice cream from a street vendor near the border, but once again we had the wrong currency. Again a surprise, the nice lady served us up ice cream at no charge. Welcome to Poland.

The currency in Poland is Zlotti's one dollar is equal to 1,000 Zlotti's.

We visited Gdansk. That city was one of the most bombed out cities during WWII. The people were proud of the way their city looked, and after the war, rebuilt it to be exactly the same as before the bombing. They used old photographs as their guide. Walking down the street one day, we passed by a butcher shop. The aroma from inside made us enter to check it out. The aroma was 'polish sausage'. We bought some of the smoked sausage, and walked down the street eating it like one would eat an ice cream cone. MMM good. Recall Gdansk is the city where Lech Walensa lived. In about 1990, Lech started the movement that would eventually topple the Soviet Union.

In the Soviet system, all property except personal items like clothing was owned by the state. Farming was done 'collectively'. Workers were assigned to each collective, and the collective was stocked with tractors and other field equipment. The yield of grain, eggs, meat, or whatever, was not the property of the workers. The yield was stored or shipped off to where it was distributed. After the harvesting machines had reaped the crop, there would be some small amount of grain that was accidentally dropped in the field. The workers were allowed to collect that fallen grain. The process of collecting was called 'Gleaning'. It was hard work. They stooped to pick up a few grains at a time, but it was THEIR GRAIN.

Collective farm

Gleaning

Another day, we were traveling down a road, and noticed that on this bright sunny day, the sky was getting smoggy. As we got into the middle of the town of Katawice, we could look up and not see the sun. The sky was completely smogged in. This was caused by another Soviet endeavor. Wherever the Soviets found a natural resource they wanted to use, they built a manufacturing plant to extract the resource, and they built apartments for the workers that would operate the plant. In this case, the natural resource was coal. A power generating plant was built over the coal field to minimize transportation costs. Now, burning coal to generate steam, and thus electricity is a dirty process, and especially so if you don't care about what it does to the atmosphere. Here is what that city looked like on a bright sunny day.

Smog

In Warsaw, we found a place to stay that was formerly a training place for speed skaters. In the Soviet days, athletes were paid to do nothing but train at their sport. The facility had an oval race track surrounded by rooms for the athletes. It was a pretty nice setup, and an excellent vantage point from which to check out Warsaw.

We had been warned about car theft, and made sure we kept the car in guarded lots as we moved about town. One Sunday, we got careless, and parked it immediately outside the post office. We were in the post office for about five minutes, and when we came out viola! The car was gone. We both stared down at the vacant space, and said, "it's gone". Now what to do? We found the local police station, and explained what had happened in sign language. They told us to come back in an hour when a translator would be available. The interpreter gave us a 'stolen car' form to fill out. Now we had a document to use in talking to the Renault people. We found a Renault distribution center, and was able to converse with the receptionist who spoke English, Polish and French. We explained what had happened. The receptionist facilitated our phone call to the Renault agency in Paris. Apparently the insurance we had was intact, and the Renault people believed our story with the help of the 'stolen car' form which we faxed to them. Four days later, a new Renault was delivered to our abode at the speed skating rink. The only thing we lost was some undeveloped film. While the car was gone, we rode the street car to wherever we wanted to go. The streetcar was a pleasant surprise. We found that when the streetcar was full, teenagers got up to give us their seats. Imagine that happening in the U S.

LaVonne developed a tooth ache resulting from a filling that fell out. We asked at the front desk where we might find a dentist. We were told that the dentist would be there the next morning. LaVonne was led into the dentist office, and noticed that the drill apparatus was a pulley driven affair that was way out dated. She was asked to sit in the chair. The dentist started to drill out the cavity without the use of Novocain. I knew it must hurt when I heard the moans coming out from inside. She said that the assistant leaned on her shoulder to keep her in place during the procedure. A wad of amalgam was put into the newly cleaned cavity. She was very happy to get out of that office. The next day, the new filling fell out. We asked if there was a European dentist nearby, and were given an address. Sure enough, the European dentist had all the right tools, and fixed the problem.

I wanted to tell about our five hour tour of Auschwitz, but I cannot. The horrors were too great. When we left Auschwitz, we felt emotionally drained.

Czechoslovakia

Czechoslovakia is now divided into two countries, Slovakia, and the Check Republic. We saw it before the division. The division was carried out peacefully without war or bloodshed. What a concept!.

As we drove down a minor road one day, we came upon three youths who tried to flag us down. They had run out of gas, and wanted help. I sensed danger there. If we stopped, we would be vulnerable to anything they wanted to do to us. We didn't stop. We then talked about it as we drove away, and decided that we were in their country, and we owed it to them to help. We turned the car around, went back, and watched as they siphoned a gallon of gas from our tank. It was all done in sign language. We felt very good about the event, and knew that we were acting as good American Ambassadors.

Yes, and I got a speeding ticket. We were going down a lesser used road, and a Cop stepped out on the side of the road with a stop sign. It was not in English, but the familiar octagon shape and red color left no doubt as to what it was. I stopped, and got out of the car. The Cop told me what the problem was in his language that I did not understand. I just stood there playing dumb, and when the speech ended, I smiled, and headed back to the car. He quickly got between the car and me and started his speech again. I listened politely, smiled and again tried to get back in the car. No way! He again stopped me, and got out a ball point pen. He pointed at the roadway, and wrote 40, and then he pointed at our car, and wrote 50. I got the message. He apparently had a radar gun, and got me as I came over the hill. I asked what the fine was in sign language, and he wrote what it was in his currency. It turned out to be $1.45 I paid it in cash, and everyone was happy.

Germany

In what was East Germany, the work was already underway to reunify, and rebuild the country. The Berlin Wall had come down only three months before we got there.

We remember visiting Germany some ten years before this trip. At that time, we passed through the Berlin wall at 'Checkpoint Charlie'. Checkpoint Charlie was manned by Soviet guards with guns. I remember approaching in our rented car, and being stopped by the guards. They wanted our passports and said, "in bring". I collected the passports and brought them inside the guard house. We were allowed to pass. The contrast between East and West Berlin at that time was very striking. The west had been built into a thriving city. In East Berlin, all of the property was owned by the state. The state didn't care much about maintaining the buildings with fresh paint. They apparently only cared about maintaining factories, and other production equipment. This was real evident as we passed out of Berlin (which was an island inside of East Germany) down one of the corridors leading to West Germany proper. We were told to stay on the autobahn until we got to West Germany. I saw a sign and an exit ramp to Bad Durensburg. What the heck! Off the autobahn we went. Again, we found a town that was in disrepair. The people there did not own their own houses, and had to rely on the government for painting and the like. We went into a café, and had a bite. We were the show! We used our West Marks to buy things. They were very happy to get the West Marks. We noticed that the people were nicely dressed. That clothing was their own!

Now, on our round the world trip, we reentered Bad Durensburg (Bad means bath in German). This Bad Durensburg was a far cry from the earlier one. All of the buildings were freshly painted, and there were flower gardens everywhere. Private ownership and the pride that comes with it had transformed the city.

We drove through Austria, the Italian Riviera, Monaco, and southern France. These are very nice places to visit. English is spoken in pretty much all tourist places. After returning the car in Paris, we boarded a plane for our round trip to Portugal. Portugal is another very fine tourist destination, with friendly people, good food, and a lot to see.

We boarded our Air France flight, and flew to Singapore, rode the train into Malaysia, and bussed to Thailand. We saw elephants doing work moving heavy loads in the Bangkok bus terminal. We saw fully dressed Muslim women swimming in the sea fully clothed. These are just amazing countries and very tourist friendly. Put them on your list!

The next flight was to Jakarta, the capital of Indonesia. Then we crossed Java slowly by bus and went on to Lombock, and Bali.

Bali is a most magical place.

This is the little house that we rented for a week. It had no air conditioning, and the daily temps were in the nineties. Each morning one of the help brought around little dishes of cooked rice. They were placed near the cabin, and about four feet off of the ground. They were offerings to the Gods. If we hung our laundry out to dry, and it was higher than four feet, we were told that it must be lowered so as not to offend the Gods. The complex had about 12 of these little cabins, and a bar/restaurant. It was situated on 'Monkey Forest Road'. If a tourist tried to book a place to stay, they would probably end up in one of the big hotels on the beach. That is not Bali.

If you walk down Monkey Forest Road, you can expect to encounter monkeys. We did, and were careful to hold on to our glasses. It seems the little guys like glasses, and will jump on your shoulder to steal them.

We stopped by a rice paddy, and watched what was going on. The farmer was preparing the bottom of the paddy for planting. He had a couple of oxen yoked to a wooden plow. The plow was a 2 x 8 strapped to a frame. He drove the team across the paddy, and shifted his weight from one side of the 2 x 8 to the other to get the bottom smooth. He saw me watching, and motioned for me to get into the paddy. I did so in bare feet, and was pleasantly surprised by a pretty solid mud bottom. We smiled at each other, and exchanged sign language (not the same as deaf mutes do, probably more like body language). I asked if I could take his picture, and got the OK. Again in sign language, I told him that we would mail the prints to him. LaVonne had a supply of envelopes, and handed one to him so he could address it. He did so. We were happy for the experience.

We were getting a bit travel weary after about 10 months on the road. Mind you, we never made a hotel reservation for the entire trip. We booked a flight out to New Caledonia, and stayed there a few days before flying back to New Zealand.

When we got back to Vonnie T, the work began. She needed a new coat of bottom paint. We also decided to refinish the masts and booms, revarnish the cabin sole, install a new radar, and new GPS (the GPS devices had come way down in price). LaVonne sanded and varnished the sole while I sanded the spars. The yard spray painted the spars with white Emron Enamel.

We launched the boat, sold the car, and gave hugs to our many friends there.

We headed north, and were in radio contact with the nets, and several friends. One of our friends had a problem in a storm and blew out his Jib. We had a stay sail that was little used, and offered it to him. We agreed to meet at Minerva reef, about 200 miles south of Fiji.

Minerva reef sticks out of the water only four or five feet at high tide. We wouldn't see it until we were less than a mile away. Our new GPS would show us where it was, but what if the chart showing its location was off by a mile or two? We sighted the reef, and sailed into the entrance, and anchored near Rebecca. We had a delightful reunion with Dave and Linda. We gave them the stay sail that could be hanked onto their forestay. They would sail to New Zealand with it, and then find a boat headed north to Fiji for the return of our stay sail. The rule of the sea among cruisers was that if someone needed something you had, you gave it to him, and if you needed something, someone would come to your aid. Rebecca departed the next day at first light, and we were left to explore the reef.

I toured the inside of the reef with the dinghy, looking for fishing spots. I had good luck for a day or two, and then a big problem arose.

Ships Log   Minerva   Reef, 23°S Latitude   179°E
                Longitude, 19 April, 1993

After my spearfishing across the 'lake' today, the dingy engine quit on me about 50 yards from the boat. No problem, I thought, that engine and I got along well. I knew all of its idiosyncrasies. I pulled the starter rope a couple of times to no avail. Then I checked the fuel lines, choked it, unchoked it, and swore at it, all to no avail. Now I was getting nervous. I was 60 yards from the Vonnie T, and being driven downwind away from it by the light trade winds. If I continued dinking with that damn motor, I would be farther away yet, with diminishing hope of returning with the dingy. LaVonne was on the stern of the boat yelling at me to "'use the oars". Of course, I didn't take them along.

It all happened very fast, and I decided my only hope was to get in the water with my swim fins, and tow the dingy back. That sounds good, but the force of the wind against the dinghy was only a little less than the force that I could apply to the dinghy's rope. I swam hard, and my heart pounded. I realized that I was getting tired, and with over exertion, I could have a heart attack. That would put Lavonne and Vonnie-T in jeopardy too as they tried to save me.
What to do? If I let go of the dinghy, I could easily swim back. That would probably mean the loss of the dinghy. If I kept pulling at maximum effort...well. I slowed my swimming effort to a pace that I could maintain for some time. I was very slowly getting closer to the boat, now about 40 yards, and making maybe a yard a minute. I yelled at LaVonne to get the polypropylene line from behind the settee, and throw it to me. The polypropylene line was the only one on board that would float. It would float straight down wind, and I could grab it, and pull myself in to safety. For some reason, she got a nylon line and threw it. It sank. I swam and yelled louder for the polypropylene line. She tied a float to the nylon line and threw it again. Of course, the nylon line sunk, and left the little float bobbing close to the stern of the boat. I swam and yelled and cursed, swam and yelled and cursed. I was near exhaustion, but I swam. I reached the stern ladder after what seemed like hours. I flopped on the deck and glared at LaVonne for not doing what I asked. She glared at me for not taking oars. Words were not necessary. We both silently went about our business knowing we would

discuss it later, when we could do so without making it a bigger personal incident than it was.

In retrospect, it was me that should have told her about the differences between polypropylene, and nylon and showed her where the polypropylene line was stored. My fault!

After rest, dinner, and reconciliation, we got a good nights sleep. Next day, off to Fiji

Fiji

Our son Clark joined us in Fiji. He was quite an accomplished SCUBA diver. In fact he rose to the grade of Dive Master and then to Instructor. We sailed to Ono Island near the 'Great Astrolabe Reef'.

In Fiji the custom is to approach each small village, and ask the Chief for permission to anchor there. We would also ask him for permission to fish. The Fijian Islanders are very friendly people. We were always welcomed. The custom was for the boat captain to bring some 'Kava Roots' to offer the Chief as a tribute. We did that at every stop. Here the Chief, whose name was Paul, greeted us. The custom was to crouch down so that we were not taller than the Chief. The custom also calls for a 'Kava Ceremony of Welcome'. As LaVonne, Clark and I sat around a Kava bowl, I offered the Chief our gift of Kava roots. He accepted the gift, and beckoned for us to join in the welcoming ceremony. The Kava had been previously prepared by mashing the roots in water. An attendant scooped out a half coconut shell of Kava brew and offered it to the Chief. He clapped his hands once and drank it down in one gulp. He then clapped twice and gave the shell back to the attendant. Now it was my turn. We were all sitted cross legged around a big Kava bowl. The Attendant approached me with a half shell filled with Kava Juice. I clapped my hands once and took the shell. Yes, I drank it down in one gulp, and then clapped twice before giving it back to the attendant. And so it went all around the bowl. Each of our party was given Kava. Then the Chief got another, and all of the rest of us were to follow.

An aside. Kava is bad tasting stuff. It is slightly narcotic. It makes your tongue slightly numb. I guess the natives drink a lot of this stuff, but not me. Two bowls were enough for us. I thanked the Chief and told him I had some boat chores to do. We had permission to anchor, and to fish in their waters. We would stay there for 7 or 8 days, and get to know the villagers, and they got to know us.

Making Kava

We all toured the Great Astrolabe Reef the next day with fins and snorkels. Wow what a wonderful reef. There were many different kinds of coral with every color of the rainbow. There were fishes galore, and steep reef faces for them to hide in.

There was a small dive resort near the village. Now we would have a source of SCUBA air. Clark went dive fishing with the SCUBA master. One day he came back to Vonnie T with a 60 pound Wahoo. He was very proud of his catch. He explained how he shot the fish, and then hung on as the powerful fish took him for a ride trying to get away. Now, what to do with that big fish?

We had enough smaller fish to last us many days. We decided that the best thing was to give the fish to the village. Clark went ashore with the fish and sought out Paul the Chief. Clark gave the fish to Paul. As it turned out, there were visiting people from another small village on the island. It was custom that if a village received a nice gift while a visiting Chief was there, the gift would be given to the visiting Chief. OK, the visiting Chief got the fish and now the onus was on him to make it right with the home Chief. The visiting Chief then gave the home Chief some money. Now everyone was happy.

Clark stayed afterward to participate in the Kava ceremony to celebrate.

We sailed back to the main island of Viti Levu where the capital city of Fiji lies.
Another nuance, at that time the Fijian government was concerned about counterfieting. The Fijian currency was brightly colored. They had instances of theives copying the currency on colored copiers. Colored copiers were now banned in Fiji.

Clark boarded an airplane for home, and we started preperations for our voyage to New Caledonia. We had been traveling abord Vonnie T and doing land travel for over six years. We talked about how much longer we would do this. Next stop would be New Caledonia, and then it was a 300 mile jump to Austrailia for the next typhoon season. Each of the new islands was not 'wow' whearas, when we started, each was 'WOW'. Maybe it was time to start thinking about selling our boat, and getting back to land life. We knew that once on the market, a boat could take a year or more to sell. We thought 'let's get the process started in New Caledonia'.

We left Fiji on a nice day with a good weather forecast. It would be three or for days to reach New Cal. During the second day out, and at night a new problem arose.

Ships log:
20°S   Latitude   171°E   Longitude,   10   July   1993

For the first time in six years I was afraid at sea. We had been uncomfortable many times, but not afraid of foundering. Today, some 100 miles east of Fiji, I saw water above the cabin sole of the boat. As usual, bad things most always happen at night. The wind was blowing 20 plus knots, and kicking up 5 to 7 foot plus seas. On inspection, I found the engine compartment near full too. We had a major leak! If we didn't find it, the cabin would fill with water, and the boat with its 10,000 pounds of lead in the keel, would sink. The cursed little automatic bilge pump either wasn't working, or was not able to keep up. LaVonne started to operate the manual pump and I began to search for the leak. The boat had 13 through hull fittings, and about 20 hoses that either brought in seawater for cooling or washing, or returned waste to the sea. I regularly checked each of those hoses, and its clamps for fear of leakage. I figured that one of those hoses was the culprit. An hour later I had checked all of the hoses without finding the leak. Lavonne kept pumping, but didn't seem to be keeping up with the leak. A low degree of panic was setting in as I started to examine every square foot of hull below the water line. Maybe we hit a big object that ruptured the hull. This wasn't likely since our fiberglass hull was about 0.4 inches thick, and very strong. We had heard no loud noise that would have meant collision with something big. My inspection ended in the bow when I opened the anchor chain locker. Every time the bow would go down on the backside of a wave, the bow would dig in slightly, and a large amount of water came gushing in through a one-inch open hole. The hole was once filled with our starboard running light.

Happy day! A problem discovered was a problem solved. Early on, I bought a set of tapered wooden plugs for the expressed purpose of plugging an unknown hole in the hull. I quickly put a couple of different sized plugs in my pocket, and started topsides with a hammer in my hand. I crawled forward on the *heaving deck to the bow. The running light cavity was only about six inches down from the deck. I reached over the side and pounded one of those pretty plugs into the hole. Back in the cabin, I relieved Lavonne from her pumping duties. Soon the bad water was back in the sea. We started the engine and saw that it had not been damaged. Whew!*

Later, I discovered that the running light housing, which was made of stainless steel, was fastened together with bronze screws. Electrolytic corrosion had once again struck. The least noble metal, bronze, had been eaten away by the current of electrolysis, which occurs when dissimilar metals are submerged in the electrolyte, seawater.

By the time we got to New Caledonian waters, we were dead tired. We anchored behind the first outer island we found. We were supposed to directly to the customs dock when entering a country but fatigue trumped what we were supposed to do. We both fell into a deep sleep only to be awoken by the French Gendarmes. They knew what we were supposed to do. They asked for our papers, and then commenced to do a thorough search of Vonnie T. They were looking for drugs, or guns.

An aside about guns: When we started out six years previously, we were concerned about protecting ourselves from pirates, thieves, and any bad guy. LaVonne bought a .357 stainless steel magnum six shooter. She took shooting lessons from a retired policeman in New Brighton. I bought a .38 caliber stainless steel revolver. We also bought a supply of ammunition. For some reason we thought that would not be enough fire power, so we also loaded on board a 12 gauge shotgun, and 303 hunting rifle. Each new country that we would enter required that we relinquish all firearms to the local police station while in their waters. The police would give the firearms back on the day of our departure. We couldn't abide with that. If we needed weapons, it would most likely be to defend ourselves while anchored in their waters. After much thought, we decided to ship the rifle and shotgun home. We also, after more thought, decided to hide the two pistols, and ammo. We had an excellent place to hide them. Vonnie T's cabin structure was a fiberglass shell overhead with spacer strips covered with Velcro. The cabin ceiling was nice naugahyde over foam rubber, and light plywood. The plywood was then fastened to the spacer strip with Velcro. The space between the fiberglass, and the plywood was about 1 $\frac{1}{2}$ inches, just enough to store a pistol. Also the chosen hiding place was directly over our stern bunks, easy to access from a sleep.

Gun philosophy: We were never big gun people. We believed that guns should be in the hands of police and military, and that simple guns could be used for hunting. The idea of having several gun toting customers roaming around a Texas supermarket was, and is abhorrent to us. The idea of anyone not police or military owning an assault rifle or the like was, and is abhorrent to us.

Now lets get to the practicality of our little gun stash. If a bad guy comes on board with intent to do us evil, we would theoretically take out our guns to defend ourselves. Now, as soon as the bad guy saw our gun, he would fire first. We would have paused before shooting, knowing that once we fired, we would be taking that person's life. Now we would be dead, and our own guns made it so. If we had no guns, the bad guy would have robbed us or whatever, but most probably would not have taken our lives.

We also question the validity of the tremendous lobbying arm of the NRA, but that's another story.

Back on board, we watched the gendarmes search, and were relieved to see they couldn't find our stash. Had they found undeclared guns, they could have confiscated our boat. They left us after telling us to raise a Q flag, and go directly to the customs dock.

We cleared customs, and began to explore New Caledonia. New Caledonia has some interesting history. The original inhabitants were Kanak, and still number about 40% of the total 250,000 population. The French arrived in 1853, and annexed New Caledonia by simply raising their flag. The French used New Caledonia as a penal colony, and placed some 30,000 French prisoners there over a 30 year time period. Nickel was discovered in 1864, and now New Caledonia is the number one Nickel producer in the world. There is a barrier reef surrounding the main Island. In fact, New Caledonia's barrier reef is larger than the Great Barrier Reef of Australia. In 1943 The Japanese had overran most to the Pacific. The U S Navy amassed a huge fleet of warships in the large and well protected bay near Numea. The U S fleet stopped the Japanese in the battle of the Coral Sea. That was the turning point in the war against Japan. The people of New Caledonia are still grateful to us for saving their Island from the Japanese. There are monuments containing the U S flag prominently displayed in the town square of Noumea.

After staying on the customs dock for three days, we anchored off, and began preparations to sell our boat.

Ships log

Nouméa, 22°S Latitude 168°E Longitude, 18 August 1993

Someone said "The two best days of boat ownership are the day you buy, and the day you sell". My love affair with Vonnie-T ended today.

She was in excellent shape, this beautiful girl of ours. Writing of what I did today brings tears to my eyes even as I sit in her bosom at the chart table.

On arrival in Numéa, we put out the word that Vonnie T was for sale. We visited a boat broker, and entertained a number of lookers that probably only were curious to see what she looked like. We even made FOR SALE signs. We placed an ad in the Sunday paper with the help of one of our French-speaking friends.

I thought it might take a year or two years to sell. Boats do not sell easy. Most are sold at a loss after too long. We would probably sail out of here to pass the typhoon season in Australia, and sail then into the North Pacific.

On Saturday, while LaVonne was off to a ladies lunch, a friend brought a prospect to see this fine lady, the Vonnie-T. They were Dr. Robert De

Malet, and his pretty wife Therese. Robert had practiced medicine for several years in France, and wanted to escape the grinding pressure of big city medicine. Here in Numéa, his patient load was only one third of that in France. They were happy people living in this island paradise. Robert had been looking for a suitable boat for some time.

Out to the boat we went in the dinghy. I saw them softly smile as they first looked from the water line to the masthead, and then from bow to stern. I circled Vonnie-T, the precious dear, for them to have a better look. We approached the port side, and mounted the wood and rope ladder that I had made. We sat in the cockpit for a while and discussed Vonnie-T's history. Then we toured topsides, and below as I described in warm terms each piece of the lady's equipment. Robert and Therese went on a second round by themselves, and returned to the cabin and promptly said, "We'll take her".

Now the emotions started as I thought "Do I really want to sell my beautiful lady?" Well yes, LaVonne and I had discussed it thoroughly, and we knew it was the right thing to do. My next thought was 'What the hell do I do now?' I had experience buying and selling real estate, so I *figured I could draft a purchase agreement. I had a hard time asking what their offer was without risking a hint that we had price flexibility. Somehow Robert said they would pay the asking price. They were worried that if they tried to bargain and the time dragged into Sunday, that we might have other offers. So I drafted up a purchase agreement using carbon paper, and listing all of what was and was not included. We all signed the dastardly document, and I got a security deposit check. We shook hands, and did French-style cheek kisses. All were very pleased. We boarded the dinghy, and I again circled Vonnie-T before returning them to the pier.*

I shoved off from the pier and headed back to that fine lady who was gently swinging on her anchor. Gentle breezes softly pushed waves caressing her waterline. On approach, I once again circled her in a salute of respect. I tied up the dinghy, and as soon as I touched the stanchion, the tears started to flow. I could scarcely see as I climbed on board. I walked the entire length of her decks apologizing for having sold her. I told her that she treated me better than some of my kids had treated me. We had sailed some 20,000 miles together. I had maintained lights at her masthead, and scrubbed the depths of her bilge. She had delivered us

*safely through numerous storms, and captain
error. Then in the end she had returned to us
more money than we had paid for her eight years
ago. I said I was sorry out loud.*

I felt really bad.

When LaVonne returned to the pier, and called on the VHF, I took
the dinghy in to pick her up. She told me that she already knew
what had happened. The Yachtie grapevine works well. We hugged
each other firmly, and cried together. We gradually got used to
reality, and started to plan for what comes next.

New Zealand was always a favorite of ours. We would go there,
rent an apartment, and plan the rest of our lives.

We found small shipping containers, and packed away our personal
things for sea shipment to New Zealand. The final payment came
after a loan went through, and we were no longer owners of the
'Sailing Yacht Vonnie T'.

Robert had agreed to take us to the airport. There was one more
thing to take care of, those guns. We never told anyone about
them. If the word got out to any authority, we would have been in
deep trouble. We couldn't take them with us on the airplane since
they were never declared when we entered New Caledonia. Now,
we sat in the cockpit discussing what to do with them. Somehow
we decided to drop LaVonne's .357 over the side. She did so with
little remorse. My gun, the .38 was brand new, and had never been
fired. We decided to offer it to Robert. I would offer to give it
to him on the morning we were to leave. If he didn't want it, I

would drop it into the sea. He did take that gun, and I hope to this day that it did not get him into any trouble.

We continued to correspond with Robert and Therese for many years. They related all of the improvements that they made on Vonnie T. Robert called her 'his expensive mistress'. They did some small voyages around the Western Pacific. One day about ten years ago Therese e mailed us to report that Robert had died. She sent us video of burying his ashes at sea from the deck of Vonnie T. There were many of the couples boating friends at sea with them in a small flotilla for the burial.

Later, she reported that Vonnie T had been sold to a New Zealand couple, and that she was now harbored in New Zealand.

Our sailing odyssey is over. In retrospect, I think that I wrote too much detail on some of the tight spots we were in. Remember, we were out for six years, and for every tight spot there were one hundred very nice experiences.

What's Next?

We returned to New Zealand and took and apartment for two months. In that time, I studied Climatological data. We wanted to live in an English speaking place that had a good climate. It kind of boiled down to southern California, the east coast of Australia or New Zealand. We came close to emigrating to NZ, but in the end we knew that it was just too far from family and friends. We flew to Los Angeles (a 13 hour flight), and drove to San Diego to visit some old friends. We listened to the radio on the trip down, and heard a variety of languages from English to Spanish, and even

Korean. The weather was near perfect and we decided right then that San Diego would be our new home.

We rented a home in Scripps Ranch, and started to build our new life. LaVonne got active in painting, and writing. I took up building and designing stage sets for the Scripps Ranch Theater. I built most of the sets for about four years. The golf bug bit me, and I became a regular golfer (albeit not a very good one).

We bought the house on 11153 Saunders Court in 1995. It was a fixer upper, and we got it for a very good price. Over the next 15 years, we would completely rebuild the house, and add 600 sq ft onto it. New windows, wiring, flooring, A/C were installed. I did the design work on the expansion, and saw the process through from Print approval by the City to supervising contractors, and doing a lot of the work myself. Here are a couple of pics of the finished product.

Cherry Wood Kitchen

Pool yard

I also took up wood working in a big way. I bought Drill Press, Router, various Sanders, Joining equipment, and tuned up the old table saw and band saw from Minnesota. I decided that staining wood was something that I wouldn't do. If I wanted honey color, I would buy good teak wood. If I wanted a light blond color, I would buy good Maple. If I wanted a deeper wood tone, I would buy good Cherry. I built bedroom sets, fireplace mantles, and a full set of office furniture. It was fun, and very satisfying.

Boat # 9 Diversion

In the summer of 2006, we took a trip down the Mississippi river for a few days with Cousins Bob and Debbie on their houseboat. We overnighted in a marina, and I did my usual thing of walking the docks and looking at boats. LaVonne and I had often talked about taking a boat from St Paul MN to New Orleans on the Mississippi. We never pulled it off because it would cost too much to rent one, and owning would pose the problem of selling at a loss. But, alas, it was October, and I knew the owner of any boat not sold before winter would be getting a bit desperate. We talked about it, and the search began. We finally found a boat in a Stillwater.

She was a 26 foot Silverton flying bridge cruiser built in 1980. Upon inspection of this 26 year old boat, we found her to be in excellent mechanical condition. The owner was an amateur race car mechanic, and happened to own a plating factory. She had twin 250 hp Chrysler engines that were very clean, and had many of the steel components chrome plated. The cabins were in need of some sprucing, but what the heck, if we got her at the right price, we would do the fix up. Negotiations began, and we got her for much less than the market value.

We took her down the St Croix to the Mississippi, and up to the Watergate Marina in St Paul. We had her hauled out, and went home to sunny San Diego for the winter. In the spring, we returned with our minivan full of fix up stuff. We spent a few weeks doing refurbishing until she looked like this:

Diversion

About the Mississippi

We toured a very good Mississippi/River Aquarium at Dubuque IA.

We learned that the Mississippi and its tributaries the Missouri, the Ohio, and others provide drainage for 2/3rds of the land area of the lower 48 states. The Mississippi has been here for Eons, and has meandered its way south, ever changing its path.

Early on, the Indigenous people were attracted to the river for water, and the rich farmlands that range from its banks. Later the Europeans came, and were attracted to it for the same reasons.

The new settlers built towns along the banks, and began to use the river for commerce. At that time there were no diesel or gasoline engines to propel watercraft, so they built prams and pushed them up river with long poles. There were no Dams, Locks, nor navigation aids.

In 1787, John Fisk first put a steam engine on a boat, and had limited success on the Delaware River. John Fulton introduced the steam boat to the Mississippi, and set the stage for rapid development of the Mississippi as a commerce center. Steam Boat captains were a special breed. Besides managing the freight and passenger business, they also had to know which parts of the river offered enough depth for transit. They dealt with the hazards of grounding, and even explosions of their steam engines.

The many steamboats carried freight, people, and, of course river gamblers. They tied all of the little settlements together. The towns were built close to the water for easy access to the river traffic. They could have been built on the higher bluffs. Flooding was not considered and so when the floods came many villages were devastated.

Before man entered the scene, the rivers took care of the excess water that came with rains by flooding laterally into natural "Flood Plains". The Flood Plains could contain vast amounts of water, and they would naturally and gradually release the water when the rains subsided.

Man changed all of that when he could not tolerate the flooding that disrupted his life near the river.

Mans response was to build Levees. A strong and well built Levee could protect a village, but it only compounded the problem for upstream villages. The Levees prevented the swollen river from spreading out to the lowlands, and caused the river, at that point, to rise to an even higher level. Many towns with Levees meant even more trouble up stream.

Steam Boats continued to be the Rivers economic engine until the advent of the Railroads. Railroad transportation of freight was much less expensive than Steam Boat.

Between 1935 and 1938 the Government constructed some 24 Lock and Dams on the upper Mississippi from Cairo Illinois to St. Paul Minnesota. Along with the dams, they built a system of navigation including buoys, and dayboards. They dredged the channel to provide for barge traffic. Today there are 27 Lock and Dams on the upper Mississippi above Cairo Illinois where the Ohio River enters. Each of these dams has an associated "Pond" immediately upstream of it. In rainy times these "Ponds" are allowed to rise to accommodate the excessive water. They accumulate the excessive water just like the Flood Plains did. They do not allow flooding. Only two of those Dams are also equipped with electrical power generation equipment. 25 of the Dams hold back only 5 to 10 feet of water, not enough hydraulic potential to make power generation feasible.

Now 'Tows' with big Tugboats ply their way along the river. They might be called 'pushes' because they do push the barges along. Below Cairo, there are no Lock and Dams and the tows may contain perhaps 20 barges. Above Cairo, the standard maximum tow is 15 barges set at 3 wide. The barges were designed so that 9 of them would fit into a lock. Thus when transiting a lock, the Tug first places the 9 front barges into the lock, and waits for them to pass. Then the Tug and the remaining 6 barges pass through, and are reunited with the first 9.

Each Tow of 15 barges carries freight equal to 90 loaded Semi Trailers.

Each Tug has the power of two railroad locomotives.

Cruising down the river takes different skills than ocean sailing. The ocean has no bends, dams, or wing dams.

The buoyage system is similar. The 'can' buoys are green and have a flat top. Going down stream, they guard the starboard (right) side of the channel. The 'nun' buoys are red, and have a conical top. Going down stream, the nuns guard the port (left) side of the channel. The buoys are moored to the bottom.

There are also 'dayboards' that are fixed to the shore, or a piling near the channel. The dayboards guide the mariner around bends in the river. They are triangular in shape, and are positioned near a river bend. When we see a dayboard, we head for it until we see another, or a set of buoys. They are green on the starboard going down river, and carry a 'mileage' marker. The mileage markers tell us how far we are upstream from the Mississippi juncture with the Ohio River. At LaCrosse WI, they read 698. At Cairo Illinois, they read zero. River mileage is kept in statute miles, as opposed to Nautical Miles (about 1.1 land miles) on the ocean.

Long ago the U.S Corps of Engineers erected 'wing dams'. Those dams are submerged cement structures built to keep the center channel in approximately the same place. They stop the river from meandering. The wing dams are perpendicular to the channel, and may be 10 to 30 feet long, and are submerged. A boat of our draft 3.5 ft. must not try to transit over a wing dam. Each wing dam is shown on our river chart, but the exact location of them is not to be trusted.

Another potential hazard, are barges pushed by a big tugboat. These may be 5 barges long, and 3 wide. They carry many tons of cargo ranging from wheat to sand. The barge 'tows' are very unmanuverable, and must be given a wide berth by little craft like ours.

And, of course, there are numerous bridges. Some are 65 ft high off of the water, but most lift, turn, or swing to allow river traffic. Each moveable bridge has a 'bridge captain' manning his bridge 24/7. We are only 17 ft to the top of our antennae, so the bridges are no problem for us.

We had envisioned being able to get off of the channel, and entering secluded coves to anchor for the night. We also wanted to anchor in front of the charming little towns along the way. Not to be. If we struck one of those submerged and unmarked wing dams, the result would have been severe damage. We were forced to go from marina to marina the whole trip, not as romantic as the anchoring we did on the Vonniet T in remote island paradises.

Nonetheless we made it to Cairo, and headed up the Ohio River to its juncture with the Cumberland. We went up the Cumberland to Lake Barkley, the first dammed up waters that constitute the TVA (Tennessee Valley Authority)

In the marina on Lake Barkley, we made a change in plans.

1. My knee was injured to the point that I had difficulty walking.
2. It was hot, and we needed to be in the air conditioned cabin, not the way to enjoy boating.
3. We knew that there were precious few marinas as we headed south.

We had had enough.

In the St Paul Marina, we became friends with a couple that showed an interest in Diversion. I phoned them, and after some negotiation, we made a sale.

Scratcher Box, U S Patent 5577287

Brother George and his wife Carol owned a financial services store. They offered check cashing, tax preparation, money orders, and other like services. One of the things that they sold there was scratch off lottery tickets. Each ticket had a secret number hidden under a coating of waxy substance. The buyer of the ticket would scratch off the waxy stuff with a coin or a fingernail revealing the secret number. Some numbers had no payoff, some a few dollars, and some paid off big money.

In George's store the customer stood on the carpet in front of the service window, and scratched off the stuff onto the carpet. It made at first a small mess, and as the day went on the mess got bigger. He wanted me to invent something that would capture the scratch off debris, and make it easy for the customer to quickly see his reward. We talked about different ways to make a device that would scratch off, and contain the debris.

I went home, and looked around my shop garage. It would have to be the place to make prototypes. The Scratcher box would have to be made of Acrylic. I bought some Acrylic sheeting, and some bending, cutting, welding and forming equipment. I taught myself how to work with Acrylic.

I thought and thought about how to do it. I experimented with many alternative ways to do it. The resulting 'Scratcher Box' is made of Acrylic with a glass scraper blade. The box has a slit in front to introduce the ticket. Inside the box is a glass blade mounted on Acrylic overhead. It is spring loaded from above and stopped so that if just touches the top of the slot. Below and inside is a table that can rotate to the rear as the ticket is introduced. As the ticket is withdrawn, the table rotates forward creating a space between the table and the glass that is just enough for the ticket to pass, but not enough for the ticket to pass with the waxy substance. The glass blade then removes the waxy substance.

Critical dimensions are the angle of attack of the glass blade, the gap between the table and the glass blade, the parallel perfection between glass blade and the table, and the force of the spring pushing the blade onto the ticket.

I made many prototypes before I was satisfied that I had an optimum design. Each prototype was tested with 'dummy' tickets that I obtained from the Lottery Commission. Then, in my garage shop, I made about forty Scratcher Boxes. Next was to place them in store locations where they would get the real test. I found about 12 stores locally that sold lottery tickets. Each store manager was eager to allow the Scratcher Box to be tested by real customers in real time. They too had had enough of the mess that the waxy substance left on their floors.

It was a lot of fun to watch customers purchase tickets, put them into the Scratcher Box, and remove the wax. They mostly got a grin on their face, and many bought more tickets to use in my box. I knew that I had a winner.

These lottery tickets were sold all over the U S. I figured that there would be well over 5,000,000 locations. If I could get half of those locations to buy a Scratcher Box that might yield a profit of, say $10, well that was a lot of money.

I went daily to the 12 locations that were testing the prototypes. Usually they were working, but sometimes the box had torn the ticket in half, and sometimes the scratch off was not complete. The managers were patient with me as I replaced the defective box with a new one. Each manager wanted the project to be successful. All of the Scratcher Boxes worked some of the time, and some of the Scratcher Boxes worked all of the time. That was not good enough. I needed all of the Boxes to work all of the time. Once that level of consistency was achieved, we would then be confident enough of the design to spend the money for Injection molds that would produce the Box components at a low cost.

And so it went for about a month. I tried to refine the design so there would be no failure, but my garage shop was incapable of reproducing the exact dimensions necessary every time. I was getting emotionally fatigued by the process. I needed a Machinist to work with me to get the necessary repeatability' I did not find such a person.

I had spent many hundreds of hours working on Scratcher Box, getting a Patent, building, testing and trying to promote it.

I abandoned the project. I did not like failure, but I could no longer proceed. I was emotionally drained.

United States Patent [19]

Olson

[11]	Patent Number:	5,577,287
[45]	Date of Patent:	Nov. 26, 1996

[54] **LOTTERY TICKET SCRAPER**

[76] Inventor: **Thomas C. Olson**, 11153 Saunders Ct., San Diego, Calif. 92131-1927

[21] Appl. No.: **618,838**

[22] Filed: **Mar. 18, 1996**

[51] Int. Cl.⁶ B44B 11/00

[52] U.S. Cl. 15/93.3; 15/93.4; 15/236.01; 30/169

[58] Field of Search 15/93.1, 93.4, 15/236.01, 102, 77; 30/164.9, 169, 280, 272

[56] **References Cited**

U.S. PATENT DOCUMENTS

4,765,842	8/1988	Sanders et al.	15/77
4,733,061	12/1988	Ritter, Jr.	15/236.01
5,255,385	10/1993	Clark	15/77

5,355,542	10/1994	Cotanese et al.	15/93.1
5,402,549	4/1995	Forrest	15/77
5,419,001	5/1995	Fox	15/236.01

Primary Examiner—David Scherbel
Assistant Examiner—Terry G. Scollon
Attorney, Agent, or Firm—John R. Duncan; Frank D. Gilliam

[57] **ABSTRACT**

A device for scraping an obscuring coating from a game ticket, such as a "scratcher" type lottery ticket. A housing has a slot sized to receive a lottery ticket. Inside the housing a scraper blade is positioned above the slot and a rotatable table below the slot. A ticket pushed into the slot will rotate the table away, then will pull the table back toward the slot when the ticket is pulled outwardly. The scraper blade is spring loaded against the ticket to scrape away the coating as the ticket moves out of the slot, without damaging the ticket. The table is adjustable to optimize operation.

18 Claims, 2 Drawing Sheets

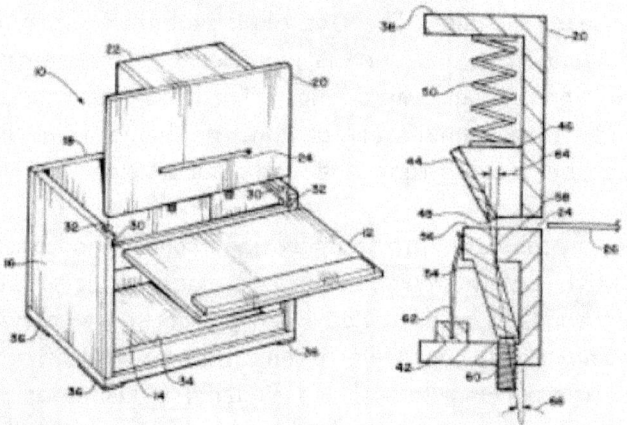

Pacemaker

After my extensive background of cardiovascular exercise in running and swimming, I thought I had a very strong heart. Well maybe it was strong, but I went through a period of frequently feeling light headed. I found that if I cut out the caffeine, and drank plenty of water the episodes decreased. However, they still were a bother, and I worried that I might pass out at an inopportune time. The Doc analyzed me, and pronounced that I had Atrial Fibrulation. My pulse rate was low, about 60 resting, and the pulse was irregular. The Doc explained that the blood flows into the Atria first, and then is pumped in to the Ventrical. The electrical signals were mixed up. The Doc referred me to a Cardiologist who immediately put me on blood thinners. After more tests, he decided that I needed a pacemaker.

I am lying on the table in the operating room. I had been given a mild sedative, but was awake. The Doc made an incision below my left color bone. The area had been previously deadened. I was talking to him about boats. He told me that he would first prepare a bed for the pacemaker with his finger. He pushed aside muscle and fat tissue, and created a little pocket. The pacemaker would be about the size of a small watch. Then he showed me the little hook like end on a wire, and the little camera that would show the way. He said, "now we are entering the vein, and proceeding to the heart. Now I see the heart, and am entering with the hook. I am now crossing the Ventricle to the other side. I am anchoring the hook in the wall of the Ventricle. There, almost done. I am withdrawing the camera, and leaving the wire." I did not feel a thing. I talked some more about boats to sooth myself. The Doc hooked up the pacemaker, sewed up the incision, and sat me up. The whole deal took less than 15 minutes, but cost about $50,000.

Now the Doc could adjust my heart rate, and did so at 80 beats per minute. There was more blood flow, and the irregular heart beat was greatly reduced. The Doc gave me a little modem to place on the headboard of my bed. The pacemaker communicated with the modem, and the modem relayed the information to the Doc's office. I felt better.

About Dogs

We have a Standard Poodle. Her name is Emma. We got her as a 10 week old puppy on LaVonne's birthday. LaVonne shared birthdays with her grandmother Emma. She thought it appropriate that our new little dog be named after grandma Emma. I agreed. LaVonne picked her out of a litter of 10 puppies that scampered around like a little happy ball of fur. On the way home in the car, Emma threw up in LaVonne's lap. She was scared. At home, things seemed OK, but at bed time we barricaded her in the kitchen so any 'accidents' could be easily cleaned up. And now the crying started. It seemed that she cried all night. She was lonesome for her siblings.

Time passed, and LaVonne housetrained her with a lot of patience. I did the petting.

Emma is now six years old, and has brought us much joy. We were always happy people, but that dog brought a new level of joy to our home. She is very intelligent, and quickly learned the commands of sit, come, stay, and etc. Every time I come home, she meets me at the door with a wagging tail. Every time either of us enters a room where she is resting, her tail starts to wag. She gives us unconditional love. I love that dog.

When Emma Dog wants something, she sits in front of me patiently looking me in the eye with those eyes that almost speak. She stays like that until I figure out what it is she wants.

Emma Dog gets two walks a day in the canyon behind our house. She is always hunting for critters. She sees a rabbit or a squirrel occasionally, but mostly enjoys the hunt. They say that a dog's nose is about 1000 times better than that of a human. Emma trots along with her nose to the ground looking for the scent of a paw that might have crossed her path. She takes care of her 'business' while in the canyon, and readily come back to us to be leashed if we go out on the street. If she sees a rabbit, that trumps any responsibility she feels for us, and the chase begins.

I think about the day, in about 10 years when she dies, or must be put down. If we make that final trip to the vet, I will hold her in my arms through the whole procedure. There will be much grief in our house. I will look for her in her favorite spot for months. I know that if I went first, she would also grieve. Maybe we can go at the same time.

I would recommend a book about dogs for you. It is 'The art of Racing in the Rain'. It is written through the voice of a dog.

Emma Dog

Travel

I have related stories of much of our travel. The traveling started with Trips to Mexico to escape the Minnesota winters. Our most recent trips have been by Airplane to China, Japan, Cambodia, and Vietnam. Here is a full list of the places we have visited:

USA

Bahamas

Jamaica

Cook Islands

Canada

Samoa

Mexico

Turks & Caicos

Tonga

Ireland

Puerto Rico (US)

Fiji

England

US Virgin Islands

New Zealand

France

British Virgin Islands

New Caledonia (Fr)

Germany

Anguilla

Costa Rica

Belgium

St Martin (Fr. & Neth.)

Viet Nam

Russia

Luxembourg

Nevis

Estonia

Latvia

Lithuania

Poland

Czechoslovakia

Austria

Hungary

Yugoslavia

Italy

Spain

Portugal

Greece

Turkey

Liechtenstein

Monaco

Morocco

Israel

Singapore

Malaysia

Thailand

Indonesia

China

Netherlands

St. Barts

Cambodia

Denmark

St Kitts

Norway

Antigua

Sweden

Montserrat

Finland

Guadeloupe   (Fr)

Martinique (Fr)

St Lucia

Dominica

St. Vincent

Carriacou

Grenada

Trinidad/ Tobago

Venezuela

Columbia

Ecuador

Peru

Chile

Argentina

Brazil

Bolivia

Paraguay

Uruguay

Dominican Rep.

Galapagos (Ecu)

Marquises (Fr)

Hawaii (US)

Tuamotos

Societe Isls (Fr)

We have never regretted the money we spent on traveling. Traveling is always an enriching experience.

During all of our travels, both LaVonne and I have always enjoyed meeting and talking with the local people. We adopted the philosophy that we never met a stranger. By the time we met someone we quickly became friends.

Some people like to keep their doors locked, and are careful to plan any trips so that they can stay at safe places.

We don't lock our doors, and rarely make hotel reservations. We have exposed ourselves to a bit more danger than most.

We have tasted the world.

A Very Peaceful Place

As I sit in our back yard watching the trees, the clouds and the blue sky, I am reminded of another page in my life that was peaceful.

We made several trips to the HF Bar Ranch in Saddlestring, Wyoming. Saddlestring is at an altitude of 5500 feet, and is about 40 miles from Sheridan Wyo. The ranch lies on a meandering creek that flows heavily from snow melt in the Tetons. The creek has a lot of very good Trout fishing holes. There are about twenty cabins along the creek, each with a deck, and an old fashioned ice box. The babbling brook is so loud, that one must talk at a slightly elevated voice to carry on a conversation. At night, that same babbling brook smoothes the way into dreamland.

The HF Bar is a dude ranch. There is a big barn there that shelters about one hundred horses. Each 'dude' that comes there is matched with a horse that will be his or hers for the entire stay. My horse was the largest one there. He was a beauty. His name was Paul. He had only one good eye.

Each day's routine was pretty much the same. Up in the morning, and walk to the mess hall after the 'dinner bell' rang. Next was a short rest period followed by a walk to the barn. At the Corral, the Wranglers had assembled all of our horses, and saddled them. We could then choose which of about 15 different trails we wanted to ride. Each ride was led by a 'Wrangler'. On those exquisite rides we crossed creeks, went through woody patches, and rode on mountain paths that were pretty narrow. We would get off occasionally to rest. One day, while crossing an open field on the 'trot', Paul put his right front foot into a gopher hole, and stumbled forward to his knees. I meanwhile, was airborne over his head. The ground came up and smacked me a good lick. I brushed myself off. Both Paul and I were OK. After the end of the morning trail ride came another short rest period before lunch in the mess hall. The afternoon ride was optional, but we always partook in it. The dinner was also optional, but we always partook in it. The food was exceptional. Pot roast, Chicken, Ribs, Fish, and plenty of vegetables were always on the menu. There was always freshly baked Apple or Blueberry pie for desert.
In the evenings, the temperatures would cool to the 50's or 60's. We build a fire in the big stone fireplace in the cabin.
Life was good. Life was easy.

When Andy was about 14, we made a vacation trip to the HF Bar Ranch. Andy wanted to work a summer at the HF Bar. He asked the head Wrangler, Dean for a job that next summer. The answer was "sure, but I hope you like to work because we have plenty of it here". At the beginning of the next summer, I took Andy to the Bus Depot for the long ride to the Ranch. We said a tearful goodbye and parted. The next day we got a phone call from him. He was sobbing (and that made me tear up too). "I am homesick Dad, and I want to go home". We talked some more, and he agreed to give it one more night, and then if he still wanted to come home, we would arrange it. Apparently he made friends with some of the Wranglers that night, and felt better about his lot there. The next day he called and said he wanted to stay. He had a very good summer riding, fishing, working and he was a pretty good eater too.

If you want to take a look at the HF Bar web site, here it is: http://www.hfbar.com/

Paul and Rider

Well, this old Dad/Husband/Engineer/Sailor/Cowboy is looking for what lies over the next mountain, and beyond the next island.

I know it will be something good.

.

www.ingramcontent.com/pod-product-compliance
Lightning Source LLC
Chambersburg PA
CBHW051857170526
45168CB00001B/145